SBAC Subject Test

Mathematics Grade 6

Student Practice Workbook

+ Two Full-Length SBAC Math Tests

Math Notion

www.MathNotion.com

SBAC Subject Test Mathematics Grade 6
Published in the United State of America By
The Math Notion
Web: WWW.MathNotion.com
Email: info@Mathnotion.com

ISBN: 978-1-63620-073-6

The Math Notion

Michael Smith has been a math instructor for over a decade now. He launched the Math Notion. Since 2006, we have devoted our time to both teaching and developing exceptional math learning materials. As a test prep company, we have worked with thousands of students. We have used the feedback of our students to develop a unique study program that can be used by students to drastically improve their math scores fast and effectively. We have more than a thousand Math learning books including:

– **SAT Math Prep**

– **ACT Math Prep**

– **SSAT/ISEE Math Prep**

– **Accuplacer Math Prep**

– **Common Core Math Prep**

–**many Math Education Workbooks, Study Guides, Practice and Exercise Books**

As an experienced Math test preparation company, we have helped many students raise their standardized test scores—and attend the colleges of their dreams: We tutor online and in person, we teach students in large groups, and we provide training materials and textbooks through our website and through Amazon.

You can contact us via email at:

info@Mathnotion.com

Get the Targeted Practice You Need to Ace the SBAC Math Test!

SBAC Subject Test Mathematics Grade 6 includes easy-to-follow instructions, helpful examples, and plenty of math practice problems to assist students to master each concept, brush up their problem-solving skills, and create confidence.

The SBAC math practice book provides numerous opportunities to evaluate basic skills along with abundant remediation and intervention activities. It is a skill that permits you to quickly master intricate information and produce better leads in less time.

Students can boost their test-taking skills by taking the book's two practice SBAC Math exams. All test questions answered and explained in detail.

Important Features of the 6th grade SBAC Math Book:

- A **complete review** of SBAC math test topics,
- Over 2,500 practice problems covering all topics tested,
- The most important concepts you need to know,
- Clear and concise, easy-to-follow sections,
- Well designed for enhanced learning and interest,
- Hands-on experience with all question types,
- **2 full-length practice tests** with detailed answer explanations,
- Cost-Effective Pricing,

Powerful math exercises to help you avoid traps and pacing yourself to beat the SBAC test. Students will gain valuable experience and raise their confidence by taking 6th grade math practice tests, learning about test structure, and gaining a deeper understanding of what is tested on the SBAC math grade 6. If ever there was a book to respond to the pressure to increase students' test scores, this is it.

WWW.MathNotion.COM

… So Much More Online!

- ✓ FREE Math Lessons
- ✓ More Math Learning Books!
- ✓ Mathematics Worksheets
- ✓ Online Math Tutors

For a PDF Version of This Book

Please Visit WWW.MathNotion.com

Contents

Chapter 1 : Review of the Whole Number Operations

Topics that you'll learn in this chapter:

- ✓ Adding Whole Numbers
- ✓ Subtracting Whole Numbers
- ✓ Multiplying Whole Numbers
- ✓ Dividing Hundreds
- ✓ Long Division by One Digit
- ✓ Division with Remainders
- ✓ Rounding Whole Numbers
- ✓ Whole Number Estimation

"Whereoer there is number, there is beauty." –Proclus

Adding Whole Numbers

Add.

1) 5,763 + 8,238 = ____

4) 2,769 + 8,872 = ____

2) 6,834 + 4,998 = ____

5) 3,196 + 2,936 = ____

3) 3,548 + 5,693 = ____

6) 7,009 + 4,992 = ____

Find the missing numbers.

7) 3,468 + ___ = 4,102

10) 631 + ___ = 2,007

8) 840 + 2,360 = ___

11) ___ + 803 = 3,945

9) 5,200 + ___ = 7,980

12) ___ + 2,156 =5,922

13) David sells gems. He finds a diamond in Istanbul and buys it for $4,795. Then, he flies to Cairo and purchases a bigger diamond for the bargain price of $9,633. How much does David spend on the two diamonds? ____________

Subtracting Whole Numbers

Subtract.

1) $\begin{array}{r} 10,512 \\ -4,411 \\ \hline \end{array}$ ______

4) $\begin{array}{r} 8,001 \\ -5,224 \\ \hline \end{array}$ ______

2) $\begin{array}{r} 5,204 \\ -3,679 \\ \hline \end{array}$ ______

5) $\begin{array}{r} 11,916 \\ -8,711 \\ \hline \end{array}$ ______

3) $\begin{array}{r} 8,520 \\ -6,483 \\ \hline \end{array}$ ______

6) $\begin{array}{r} 5,005 \\ -2,008 \\ \hline \end{array}$ ______

Find the missing number.

7) 5,263 – ___ = 2,367

10) 6,511 – ___ = 3,759

8) 7,198 – ___ = 4,742

11)7,003 – 5,489 = ___

9) 8,928 –3,764 = ___

12) 8,800– 5,995 = ___

13) Jackson had $7,189 invested in the stock market until he lost $3,793 on those investments. How much money does he have in the stock market now?

____

Multiplying Whole Numbers

Find the answers.

1) $2{,}200 \times 31$

2) $3{,}200 \times 22$

3) $5{,}790 \times 5$

4) $5{,}220 \times 3$

5) $6{,}911 \times 3$

6) $1{,}998 \times 40$

7) $2{,}893 \times 5.5$

8) $2{,}254 \times 3.5$

9) $4{,}372 \times 4.8$

10) $3{,}984 \times 2.75$

11) $4{,}900 \times 2.5$

12) $8{,}200 \times 4.5$

Dividing Hundreds

Find answers.

1) 4,440 ÷ 400
2) 1,600 ÷ 40
3) 9,990 ÷ 90
4) 4, 200 ÷ 60
5) 6, 400 ÷ 8,000
6) 2,700 ÷ 30
7) 3, 333 ÷ 30
8) 558 ÷ 45
9) 2,278 ÷ 85
10) 1,683 ÷ 55
11) 1, 582 ÷ 35
12) 9, 000 ÷ 600
13) 1,000 ÷ 2,500
14) 44.8 ÷ 20
15) 6,800 ÷ 400
16) 1,500 ÷ 5,000
17) 36.60 ÷ 120
18) 7,700 ÷ 700
19) 5,400 ÷ 600
20) 8, 000 ÷ 160
21) 18,000÷ 9,000
22) 42,000÷ 30
23) 480÷ 40
24) 63,000÷ 900

Long Division by Two Digits

Find the quotient.

1) $18\overline{)576}$
2) $14\overline{)952}$
3) $21\overline{)588}$
4) $23\overline{)299}$
5) $44\overline{)748}$
6) $26\overline{)234}$
7) $16\overline{)496}$
8) $29\overline{)1,479}$
9) $54\overline{)1,080}$
10) $41\overline{)1,476}$
11) $53\overline{)2,491}$
12) $60\overline{)2,880}$
13) $32\overline{)2,912}$
14) $77\overline{)8,393}$
15) $85\overline{)3,740}$
16) $57\overline{)4,617}$
17) $50\overline{)9,200}$
18) $25\overline{)15,400}$

Division with Remainders

Find the quotient with remainder.

1) $14\overline{)715}$
2) $16\overline{)2,750}$
3) $27\overline{)4,603}$
4) $58\overline{)2,554}$
5) $42\overline{)7,732}$
6) $63\overline{)6,737}$
7) $71\overline{)9,036}$
8) $65\overline{)8,624}$
9) $35\overline{)5,705}$
10) $92\overline{)13,161}$
11) $46\overline{)12,214}$
12) $69\overline{)42,482}$
13) $85\overline{)6,858}$
14) $87\overline{)34,304}$

Rounding Whole Numbers

Round each number to the underlined place value.

1) 7,533

2) 9,374

3) 8,883

4) 2,368

5) 5,577

6) 3,381

7) 3,520

8) 9,338

9) 8.581

10) 33.57

11) 51.69

12) 22.138

13) 6,758

14) 11,557

15) 8,838

16) 5.889

17) 1.860

18) 25.070

19) 9.332

20) 49.48

21) 28.89

22) 24,377

23) 52,158

24) 13,883

25) 9,609

26) 17,451

27) 18,768

Whole Number Estimation

Estimate the sum by rounding each added to the nearest ten.

1) 875 + 325

2) 985 + 1,452

3) 2,424 + 4,128

4) 1,576 + 6,279

5) 1,247 + 3,863

6) 6,746 + 5,121

7) 3,924 + 6,456

8) 1,785 + 7,164

9) 1,458 + 2,442

10) 5,689 + 4,151

11) 8,259 + 4,754

12) 6,788 + 3,954

13) 9,123 + 4,455

14) 6,680 + 5,358

15) 3,165 + 7,124

16) 8,859 + 6,452

Answers of Worksheets

Adding Whole Numbers

1) 14,001
2) 11,832
3) 9,241
4) 11,641
5) 6,132
6) 12,001
7) 634
8) 3,200
9) 2,780
10) 1,376
11) 3,142
12) 3,766
13) $14,428

Subtracting Whole Numbers

1) 6,101
2) 1,525
3) 2,037
4) 2,777
5) 3,205
6) 2,997
7) 2,896
8) 2,456
9) 5,164
10) 2,752
11) 1,514
12) 2,805
13) 3,396

Multiplying Whole Numbers

1) 68,200
2) 70,400
3) 28,950
4) 15,660
5) 20,733
6) 79,920
7) 15,911.5
8) 7,889
9) 20,985.6
10) 10,956
11) 12,250
12) 36,900

Dividing Hundreds

1) 11.1
2) 40
3) 111
4) 70
5) 0.8
6) 90
7) 111.1
8) 12.4
9) 26.8
10) 30.6
11) 45.2
12) 15
13) 0.4
14) 2.24
15) 17
16) 0.3
17) 0.305
18) 11
19) 9
20) 50
21) 2
22) 1,400
23) 12
24) 70

Long Division by Two Digits

1) 32
2) 68
3) 28
4) 13
5) 17
6) 9
7) 31
8) 51
9) 20
10) 36
11) 47
12) 48
13) 91
14) 109
15) 44
16) 81
17) 184
18) 616

Division with Remainders

1) 51 R1
2) 171 R14
3) 170 R13
4) 44 R2
5) 184 R4
6) 106 R59
7) 127 R19
8) 132 R44
9) 163 R0
10) 143 R5
11) 265 R24
12) 615 R47
13) 80 R58
14) 394 R26

Rounding Whole Numbers

1) 7,500
2) 9,400
3) 8,880
4) 2,370
5) 5,580
6) 3,380
7) 3,500
8) 9,340
9) 8.60
10) 33.60
11) 51.70
12) 22.100
13) 7,000
14) 11,560
15) 8,840
16) 5.900
17) 1.900
18) 25.100
19) 9.000
20) 49.50
21) 28.90
22) 24,380
23) 52,160
24) 13,880
25) 9,600
26) 17,450
27) 18,800

Whole Number Estimation

1) 1,200
2) 2,440
3) 6,550
4) 7,860
5) 5,110
6) 11,870
7) 10,380
8) 8,950
9) 3,900
10) 9,840
11) 13,010
12) 10,740
13) 13,580
14) 12,040
15) 10,290
16) 15,310

Chapter 2 :

Integers and Number Theory

Topics that you will practice in this chapter:

- ✓ Adding and Subtracting Integers
- ✓ Multiplying and Dividing Integers
- ✓ Order of Operations
- ✓ Ordering Integers and Numbers
- ✓ Integers and Absolute Value
- ✓ Factoring Numbers
- ✓ Prime Factorization
- ✓ Divisibility Rules
- ✓ Greatest Common Factor (GCF)
- ✓ Least Common Multiple (LCM)

"In order to gain the most, you have to know how to convert Negatives to Positives."

–Stubborn Clown

Adding and Subtracting Integers

Find each sum.

1) $14 + (-6) =$

2) $(-13) + (-20) =$

3) $5 + (-28) =$

4) $50 + (-12) =$

5) $(-7) + (-15) + 3 =$

6) $30 + (-14) + 8 =$

7) $40 + (-10) + (-14) + 17 =$

8) $(-15) + (-20) + 13 + 35 =$

9) $40 + (-20) + (38 - 29) =$

10) $28 + (-12) + (30 - 12) =$

Find each difference.

11) $(-18) - (-7) =$

12) $25 - (-14) =$

13) $(-20) - 36 =$

14) $34 - (-19) =$

15) $51 - (30 - 21) =$

16) $17 - (5) - (-24) =$

17) $(35 + 20) - (-46) =$

18) $48 - 16 - (-8) =$

19) $62 - (28 + 17) - (-15) =$

20) $58 - (-23) - (-31) =$

21) $19 - (-8) - (-13) =$

22) $(19 - 24) - (-14) =$

23) $27 - 33 - (-21) =$

24) $58 - (32 + 24) - (-9) =$

25) $36 - (-30) + (-17) =$

26) $27 - (-42) + (-31) =$

Multiplying and Dividing Integers

Find each product.

1) $(-9) \times (-5) =$

2) $(-3) \times 9 =$

3) $8 \times (-12) =$

4) $(-7) \times (-20) =$

5) $(-3) \times (-5) \times 6 =$

6) $(14 - 3) \times (-8) =$

7) $12 \times (-9) \times (-3) =$

8) $(140 + 10) \times (-2) =$

9) $10 \times (-12 + 8) \times 3 =$

10) $(-8) \times (-5) \times (-10) =$

Find each quotient.

11) $42 \div (-7) =$

12) $(-48) \div (-6) =$

13) $(-40) \div (-8) =$

14) $54 \div (-2) =$

15) $152 \div 19 =$

16) $(-144) \div (-12) =$

17) $180 \div (-10) =$

18) $(-312) \div (-12) =$

19) $221 \div (-13) =$

20) $(-126) \div (6) =$

21) $(-161) \div (-7) =$

22) $-266 \div (-14) =$

23) $(-120) \div (-4) =$

24) $270 \div (-18) =$

25) $(-208) \div (-8) =$

26) $(135) \div (-15) =$

Order of Operations

Evaluate each expression.

1) $7 + (5 \times 4) =$

2) $14 - (3 \times 6) =$

3) $(19 \times 4) + 16 =$

4) $(16 - 7) - (8 \times 2) =$

5) $27 + (18 \div 3) =$

6) $(18 \times 8) \div 6 =$

7) $(32 \div 4) \times (-2) =$

8) $(9 \times 4) + (32 - 18) =$

9) $24 + (4 \times 3) + 7 =$

10) $(36 \times 3) \div (2 + 2) =$

11) $(-7) + (12 \times 3) + 11 =$

12) $(8 \times 5) - (24 \div 6) =$

13) $(7 \times 6 \div 3) - (12 + 9) =$

14) $(13 + 5 - 14) \times 3 - 2 =$

15) $(20 - 14 + 30) \times (64 \div 4) =$

16) $32 + \left(28 - (36 \div 9)\right) =$

17) $(7 + 6 - 4 - 7) + (15 \div 5) =$

18) $(85 - 20) + (20 - 18 + 7) =$

19) $(20 \times 2) + (14 \times 3) - 22 =$

20) $18 + 5 - (30 \times 3) + 20 =$

21) $(\frac{7}{5-1}) \times (2 + 6) \times 2$

22) $20 \div (4 - (10 - 8))$

Ordering Integers and Numbers

Order each set of integers from least to greatest.

1) $8, -10, -5, -3, 4$ ___, ___, ___, ___, ___, ___

2) $-10, -18, 6, 14, 27$ ___, ___, ___, ___, ___, ___

3) $15, -8, -21, 21, -23$ ___, ___, ___, ___, ___, ___

4) $-14, -40, 23, -12, 47$ ___, ___, ___, ___, ___, ___

5) $59, -54, 32, -57, 36$ ___, ___, ___, ___, ___, ___

6) $68, 26, -19, 47, -34$ ___, ___, ___, ___, ___, ___

Order each set of integers from greatest to least.

7) $18, 36, -16, -18, -10$ ___, ___, ___, ___, ___, ___

8) $27, 34, -12, -24, 94$ ___, ___, ___, ___, ___, ___

9) $50, -21, -13, 42, -2$ ___, ___, ___, ___, ___, ___

10) $37, 46, -20, -16, 86$ ___, ___, ___, ___, ___, ___

11) $-18, 88, -26, -59, 75$ ___, ___, ___, ___, ___, ___

12) $-65, -30, -25, 3, 14$ ___, ___, ___, ___, ___, ___

Integers and Absolute Value

Write absolute value of each number.

1) $|-2| =$

2) $|-27| =$

3) $|-20| =$

4) $|14| =$

5) $|6| =$

6) $|-55| =$

7) $|16| =$

8) $|2| =$

9) $|54| =$

10) $|-4| =$

11) $|-11|$

12) $|88| =$

13) $|0| =$

14) $|79| =$

15) $|-32| =$

16) $|-17| =$

17) $|42| =$

18) $|-46| =$

19) $|1| =$

20) $|-40| =$

Evaluate the value.

21) $|-5| - \frac{|-21|}{7} =$

22) $14 - |3 - 15| - |-4| =$

23) $\frac{|-32|}{4} \times |-4| =$

24) $\frac{|7 \times (-3)|}{7} \times \frac{|-19|}{3} =$

25) $|4 \times (-5)| + \frac{|-40|}{5} =$

26) $\frac{|-45|}{9} \times \frac{|-24|}{12} =$

27) $|-12 + 8| \times \frac{|-7 \times 7|}{7}$

28) $\frac{|-11 \times 2|}{4} \times |-16| =$

Factoring Numbers

List all positive factors of each number.

1) 12
2) 16
3) 28
4) 34
5) 95
6) 56
7) 65
8) 70
9) 25
10) 48
11) 27
12) 63
13) 72
14) 15
15) 80

List the prime factorization for each number.

16) 10
17) 26
18) 20
19) 30
20) 40
21) 44
22) 55
23) 78
24) 96

Prime Factorization

Factor the following numbers to their prime factors.

1) 6
2) 49
3) 60
4) 4
5) 46
6) 57
7) 54
8) 38
9) 58
10) 62
11) 75
12) 88
13) 93
14) 100
15) 68
16) 90
17) 69
18) 76
19) 86
20) 92
21) 99
22) 77
23) 90
24) 74

Divisibility Rules

Use the divisibility rules to underline the factors of the number.

1) 8 2 3 4 5 6 7 8 9 10

2) 18 2 3 4 5 6 7 8 9 10

3) 55 2 3 4 5 6 7 8 9 10

4) 45 2 3 4 5 6 7 8 9 10

5) 20 2 3 4 5 6 7 8 9 10

6) 9 2 3 4 5 6 7 8 9 10

7) 21 2 3 4 5 6 7 8 9 10

8) 28 2 3 4 5 6 7 8 9 10

9) 36 2 3 4 5 6 7 8 9 10

10) 40 2 3 4 5 6 7 8 9 10

11) 39 2 3 4 5 6 7 8 9 10

12) 51 2 3 4 5 6 7 8 9 10

Greatest Common Factor

Find the GCF for each number pair.

1) 6, 2

2) 4, 5

3) 3, 12

4) 7, 3

5) 5, 10

6) 8, 48

7) 6, 18

8) 9, 15

9) 12, 18

10) 4, 36

11) 6, 10

12) 28, 52

13) 25, 10

14) 22, 24

15) 9, 54

16) 8, 54

17) 42, 14

18) 16, 40

19) 9,2, 3

20) 5, 15, 10

21) 7, 9, 2

22) 16, 64

23) 30, 48

24) 36, 63

Least Common Multiple

Find the LCM for each number pair.

1) 6, 9

2) 15, 45

3) 16, 40

4) 12, 36

5) 18, 27

6) 14, 42

7) 6, 30

8) 8, 56

9) 7, 21

10) 8, 20

11) 15, 25

12) 7, 9

13) 4, 11

14) 8, 28

15) 28, 56

16) 40, 50

17) 12, 13

18) 22, 11

19) 36, 20

20) 15, 35

21) 18, 81

22) 30, 54

23) 18,45

24) 75, 25

Answers of Worksheets

Adding and Subtracting Integers

1) 8
2) −33
3) −23
4) 38
5) −19
6) 24
7) 33
8) 13
9) 29
10) 34
11) −11
12) 39
13) −56
14) 53
15) 42
16) 36
17) 101
18) 40
19) 32
20) 112
21) 40
22) 9
23) 15
24) 11
25) 49
26) 38

Multiplying and Dividing Integers

1) 45
2) −27
3) −96
4) 140
5) 90
6) −88
7) 324
8) −300
9) −120
10) −400
11) −6
12) 8
13) 5
14) −27
15) 8
16) 12
17) −18
18) 26
19) −17
20) −21
21) 23
22) 19
23) 30
24) −15
25) 26
26) −9

Order of Operations

1) 27
2) −4
3) 92
4) −7
5) 33
6) 24
7) −16
8) 50
9) 43
10) 27
11) 40
12) 36
13) −7
14) 10
15) 576
16) 56
17) 5
18) 74
19) 60
20) −47
21) 28
22) 10

Ordering Integers and Numbers

1) −10, −5, −3, 4, 8
2) −18, −10, 6, 14, 27
3) −23, −21, −8, 15, 21
4) −40, −14, −12, 23, 47
5) −57, −54, 32, 36, 59
6) −34, −19, 26, 47, 68
7) 36, 18, −10, −16, −18
8) 94, 34, 27, −12, −24
9) 50, 42, −2, −13, −21
10) 86, 46, 37, −16, −20
11) 88, 75, −18, −26, −59
12) 14, 3, −25, −30, −65

Integers and Absolute Value

1) 2
2) 27
3) 20
4) 14
5) 6
6) 55
7) 16
8) 2
9) 54
10) 4
11) 11
12) 88
13) 0
14) 79
15) 32
16) 17
17) 42
18) 46
19) 1
20) 40
21) 2
22) -2
23) 32
24) 19
25) 28
26) 10
27) 28
28) 88

Factoring Numbers

1) 1, 2, 3, 4, 6, 12
2) 1, 2, 4, 8 ,16
3) 1, 2, 4, 7, 14, 28
4) 1, 2, 17, 34
5) 1, 5, 19, 95
6) 1, 2, 4, 7, 8, 14, 28, 56
7) 1, 5, 13, 65
8) 1, 2, 5, 7, 10, 14, 35, 70
9) 1, 5, 25
10) 1, 2, 3, 4, 6, 8, 12, 16, 24, 48
11) 1, 3, 9, 27
12) 1, 3, 7, 9, 21, 63
13) 1, 2, 3, 4, 6, 8, 9, 12, 18, 24, 36, 72
14) 1, 3, 5, 15
15) 1, 2, 4, 5, 8, 10, 16, 20, 40, 80
16) 2×5
17) 2×13
18) $2 \times 2 \times 5$
19) $2 \times 3 \times 5$
20) $2 \times 2 \times 2 \times 5$
21) $2 \times 2 \times 11$
22) 5×11
23) $2 \times 3 \times 13$
24) $2 \times 2 \times 2 \times 2 \times 2 \times 3$

Prime Factorization

1) 2. 3
2) 7. 7
3) 2. 2. 3. 5
4) 2. 2
5) 2. 23
6) 3. 19
7) 2. 3. 3. 3
8) 2. 19
9) 2. 29
10) 2. 31
11) 3. 5. 5
12) 2. 2. 2. 11
13) 3. 31
14) 2. 2. 5. 5
15) 2. 2. 17
16) 2. 3. 3. 5
17) 3. 23
18) 2. 2. 19
19) 2. 43
20) 2. 2. 23
21) 3. 3. 11
22) 7. 11
23) 2. 3. 3. 5
24) 2. 37

Divisibility Rules

1) 8   **2** 3 **4** 5 6 7 **8** 9 10

2) 18   **2** **3** 4 5 **6** 7 8 **9** 10

3) 55   2 3 4 **5** 6 7 8 9 10

4) 45   2 **3** 4 **5** 6 7 8 **9** 10

5) 20   **2** 3 **4** **5** 6 7 8 9 **10**

6) 9   2 **3** 4 5 6 7 8 **9** 10

7) 21   2 **3** 4 5 6 **7** 8 9 10

8) 28   **2** 3 **4** 5 6 **7** 8 9 10

9) 36   **2** **3** **4** 5 **6** 7 8 **9** 10

10) 40   **2** 3 **4** **5** 6 7 **8** 9 **10**

11) 39   2 **3** 4 5 6 7 8 9 10

12) 51   2 **3** 4 5 6 7 8 9 10

Greatest Common Factor

1) 2	7) 6	13) 5	19) 1
2) 1	8) 3	14) 2	20) 5
3) 3	9) 6	15) 9	21) 1
4) 1	10) 4	16) 2	22) 16
5) 5	11) 2	17) 14	23) 6
6) 8	12) 4	18) 8	24) 9

Least Common Multiple

1) 18	7) 30	13) 44	19) 180
2) 45	8) 56	14) 56	20) 105
3) 80	9) 21	15) 56	21) 162
4) 36	10) 40	16) 200	22) 270
5) 54	11) 75	17) 156	23) 90
6) 42	12) 63	18) 22	24) 75

Chapter 3 :
Fractions

Topics that you will practice in this chapter:

- ✓ Simplifying Fractions
- ✓ Adding and Subtracting Fractions
- ✓ Multiplying and Dividing Fractions
- ✓ Adding and Subtract Mixed Numbers
- ✓ Multiplying and Dividing Mixed Numbers

"A Man is like a fraction whose numerator is what he is and whose denominator is what he thinks of himself. The larger the denominator, the smaller the fraction." –Tolstoy

Simplifying Fractions

Simplify each fraction to its lowest terms.

1) $\frac{5}{10} =$

2) $\frac{28}{35} =$

3) $\frac{27}{36} =$

4) $\frac{40}{80} =$

5) $\frac{14}{56} =$

6) $\frac{32}{48} =$

7) $\frac{52}{65} =$

8) $\frac{15}{60} =$

9) $\frac{80}{160} =$

10) $\frac{55}{77} =$

11) $\frac{28}{112} =$

12) $\frac{32}{64} =$

13) $\frac{63}{72} =$

14) $\frac{81}{90} =$

15) $\frac{35}{105} =$

16) $\frac{25}{70} =$

17) $\frac{80}{280} =$

18) $\frac{12}{81} =$

19) $\frac{36}{186} =$

20) $\frac{240}{540} =$

21) $\frac{70}{560} =$

Find the answer for each problem.

22) Which of the following fractions equal to $\frac{3}{4}$? ____

A. $\frac{60}{90}$ B. $\frac{43}{104}$ C. $\frac{48}{64}$ D. $\frac{150}{300}$

23) Which of the following fractions equal to $\frac{5}{8}$? ____

A. $\frac{125}{200}$ B. $\frac{115}{200}$ C. $\frac{50}{100}$ D. $\frac{30}{90}$

24) Which of the following fractions equal to $\frac{3}{7}$? ____

A. $\frac{58}{116}$ B. $\frac{54}{126}$ C. $\frac{270}{167}$ D. $\frac{42}{63}$

Adding and Subtracting Fractions

Find the sum.

1) $\frac{5}{9}+\frac{4}{9}=$

2) $\frac{1}{2}+\frac{1}{7}=$

3) $\frac{3}{8}+\frac{1}{4}=$

4) $\frac{3}{5}+\frac{1}{2}=$

5) $\frac{1}{4}+\frac{3}{5}=$

6) $\frac{7}{8}+\frac{3}{8}=$

7) $\frac{1}{2}+\frac{7}{10}=$

8) $\frac{2}{5}+\frac{2}{3}=$

9) $\frac{5}{7}+\frac{2}{3}=$

10) $\frac{7}{12}+\frac{3}{4}=$

11) $\frac{5}{6}+\frac{2}{5}=$

12) $\frac{1}{12}+\frac{2}{3}=$

Find the difference.

13) $\frac{1}{3}-\frac{1}{6}=$

14) $\frac{3}{4}-\frac{1}{8}=$

15) $\frac{1}{2}-\frac{1}{3}=$

16) $\frac{1}{4}-\frac{1}{5}=$

17) $\frac{5}{8}-\frac{2}{3}=$

18) $\frac{1}{4}-\frac{1}{7}=$

19) $\frac{5}{6}-\frac{1}{9}=$

20) $\frac{3}{4}-\frac{1}{6}=$

21) $\frac{7}{8}-\frac{1}{12}=$

22) $\frac{8}{15}-\frac{3}{5}=$

23) $\frac{3}{12}-\frac{1}{14}=$

24) $\frac{10}{13}-\frac{7}{26}=$

25) $\frac{6}{7}-\frac{3}{4}=$

26) $\frac{4}{5}-\frac{1}{8}=$

27) $\frac{4}{7}-\frac{2}{35}=$

28) $\frac{9}{16}-\frac{2}{8}=$

29) $\frac{8}{9}-\frac{7}{18}=$

30) $\frac{1}{2}-\frac{4}{9}=$

Multiplying and Dividing Fractions

Find the value of each expression in lowest terms.

1) $\frac{1}{5} \times \frac{15}{5} =$

2) $\frac{9}{12} \times \frac{4}{9} =$

3) $\frac{1}{16} \times \frac{8}{10} =$

4) $\frac{1}{24} \times \frac{8}{10} =$

5) $\frac{1}{5} \times \frac{1}{4} =$

6) $\frac{7}{9} \times \frac{1}{7} =$

7) $\frac{6}{7} \times \frac{1}{3} =$

8) $\frac{2}{8} \times \frac{2}{8} =$

9) $\frac{5}{8} \times \frac{3}{5} =$

10) $\frac{4}{7} \times \frac{1}{8} =$

11) $\frac{7}{15} \times \frac{5}{7} =$

12) $\frac{3}{10} \times \frac{5}{9} =$

Find the value of each expression in lowest terms.

13) $\frac{1}{4} \div \frac{1}{8} =$

14) $\frac{1}{10} \div \frac{1}{5} =$

15) $\frac{3}{4} \div \frac{1}{5} =$

16) $\frac{1}{3} \div \frac{5}{6} =$

17) $\frac{1}{7} \div \frac{8}{42} =$

18) $\frac{3}{4} \div \frac{1}{6} =$

19) $\frac{2}{7} \div \frac{7}{13} =$

20) $\frac{1}{24} \div \frac{3}{16} =$

21) $\frac{7}{12} \div \frac{5}{6} =$

22) $\frac{22}{18} \div \frac{11}{9} =$

23) $\frac{9}{35} \div \frac{3}{7} =$

24) $\frac{2}{7} \div \frac{8}{21} =$

25) $\frac{1}{9} \div \frac{2}{5} =$

26) $\frac{5}{12} \div \frac{3}{5} =$

27) $\frac{3}{20} \div \frac{1}{6} =$

28) $\frac{8}{20} \div \frac{3}{4} =$

29) $\frac{5}{6} \div \frac{2}{9} =$

30) $\frac{5}{11} \div \frac{3}{4} =$

Adding and Subtracting Mixed Numbers

Find the sum.

1) $3\frac{1}{3} + 2\frac{1}{6} =$

2) $4\frac{1}{2} + 3\frac{1}{2} =$

3) $3\frac{3}{8} + 1\frac{1}{8} =$

4) $2\frac{1}{4} + 2\frac{1}{3} =$

5) $3\frac{5}{6} + 2\frac{7}{12} =$

6) $5\frac{4}{15} + 3\frac{3}{5} =$

7) $2\frac{1}{3} + 4\frac{3}{7} =$

8) $3\frac{1}{2} + 4\frac{2}{5} =$

9) $5\frac{2}{5} + 6\frac{3}{7} =$

10) $8\frac{5}{16} + 6\frac{1}{12} =$

Find the difference.

11) $3\frac{1}{4} - 1\frac{3}{4} =$

12) $6\frac{3}{5} - 4\frac{2}{5} =$

13) $4\frac{1}{3} - 3\frac{1}{9} =$

14) $7\frac{1}{7} - 5\frac{1}{2} =$

15) $5\frac{1}{3} - 2\frac{1}{12} =$

16) $8\frac{1}{5} - 4\frac{1}{3} =$

17) $9\frac{1}{4} - 6\frac{1}{8} =$

18) $11\frac{7}{15} - 8\frac{3}{5} =$

19) $14\frac{5}{6} - 11\frac{3}{5} =$

20) $18\frac{2}{7} - 14\frac{1}{5} =$

21) $9\frac{1}{3} - 4\frac{1}{4} =$

22) $6\frac{1}{8} - 4\frac{1}{16} =$

23) $19\frac{3}{8} - 15\frac{1}{3} =$

24) $11\frac{1}{9} - 8\frac{1}{8} =$

25) $17\frac{1}{7} - 11\frac{1}{5} =$

26) $16\frac{2}{9} - 9\frac{5}{7} =$

Multiplying and Dividing Mixed Numbers

Find the product.

1) $5\frac{1}{2} \times 2\frac{1}{4} =$

2) $5\frac{1}{3} \times 4\frac{1}{3} =$

3) $5\frac{3}{4} \times 6\frac{1}{4} =$

4) $3\frac{1}{3} \times 2\frac{3}{5} =$

5) $4\frac{8}{10} \times 1\frac{1}{24} =$

6) $6\frac{2}{7} \times 1\frac{1}{11} =$

7) $8\frac{2}{3} \times 3\frac{1}{2} =$

8) $3\frac{4}{7} \times 2\frac{1}{5} =$

9) $5\frac{2}{8} \times 4\frac{1}{6} =$

10) $7\frac{3}{3} \times 1\frac{3}{8} =$

Find the quotient.

11) $2\frac{2}{5} \div 4\frac{1}{5} =$

12) $4\frac{1}{6} \div 3\frac{1}{3} =$

13) $6\frac{1}{3} \div 1\frac{1}{2} =$

14) $7\frac{1}{10} \div 2\frac{2}{5} =$

15) $3\frac{1}{3} \div 1\frac{1}{9} =$

16) $1\frac{1}{10} \div 4\frac{1}{2} =$

17) $1\frac{3}{16} \div 5\frac{1}{4} =$

18) $4\frac{1}{3} \div 4\frac{3}{4} =$

19) $9\frac{1}{3} \div 2\frac{1}{4} =$

20) $15\frac{1}{3} \div 5\frac{1}{2} =$

21) $4\frac{1}{6} \div 1\frac{1}{5} =$

22) $1\frac{1}{18} \div 1\frac{2}{9} =$

23) $4\frac{2}{7} \div 1\frac{3}{10} =$

24) $7\frac{1}{3} \div 2\frac{2}{11} =$

25) $8\frac{2}{5} \div 1\frac{1}{6} =$

26) $9\frac{1}{3} \div 2\frac{1}{7} =$

Answers of Worksheets

Simplifying Fractions

1) $\frac{1}{2}$
2) $\frac{4}{5}$
3) $\frac{3}{4}$
4) $\frac{1}{2}$
5) $\frac{1}{4}$
6) $\frac{2}{3}$
7) $\frac{4}{5}$
8) $\frac{1}{4}$
9) $\frac{1}{2}$
10) $\frac{5}{7}$
11) $\frac{1}{4}$
12) $\frac{1}{2}$
13) $\frac{7}{8}$
14) $\frac{9}{10}$
15) $\frac{1}{3}$
16) $\frac{5}{14}$
17) $\frac{2}{7}$
18) $\frac{4}{27}$
19) $\frac{6}{31}$
20) $\frac{4}{9}$
21) $\frac{1}{8}$
22) C
23) A
24) B

Adding and Subtracting Fractions

1) $\frac{9}{9} = 1$
2) $\frac{9}{14}$
3) $\frac{5}{8}$
4) $1\frac{1}{10}$
5) $\frac{17}{20}$
6) $1\frac{1}{4}$
7) $1\frac{1}{5}$
8) $1\frac{1}{15}$
9) $1\frac{8}{21}$
10) $1\frac{1}{3}$
11) $1\frac{7}{30}$
12) $\frac{3}{4}$
13) $\frac{1}{6}$
14) $\frac{5}{8}$
15) $\frac{1}{6}$
16) $\frac{1}{20}$
17) $-\frac{1}{24}$
18) $\frac{3}{28}$
19) $\frac{13}{18}$
20) $\frac{7}{12}$
21) $\frac{19}{24}$
22) $-\frac{1}{15}$
23) $\frac{5}{28}$
24) $\frac{1}{2}$
25) $\frac{3}{28}$
26) $\frac{27}{40}$
27) $\frac{18}{35}$
28) $\frac{5}{16}$
29) $\frac{1}{2}$
30) $\frac{1}{18}$

Multiplying and Dividing Fractions

1) $\frac{3}{5}$
2) $\frac{1}{3}$
3) $\frac{1}{20}$
4) $\frac{1}{30}$
5) $\frac{1}{20}$
6) $\frac{1}{9}$
7) $\frac{2}{7}$
8) $\frac{1}{16}$
9) $\frac{3}{8}$
10) $\frac{1}{14}$
11) $\frac{1}{3}$
12) $\frac{1}{6}$
13) 2
14) $\frac{1}{2}$
15) $3\frac{3}{4}$
16) $\frac{2}{5}$

17) $\frac{3}{4}$
18) $4\frac{1}{2}$
19) $\frac{26}{49}$
20) $\frac{2}{9}$
21) $\frac{7}{10}$
22) 1
23) $\frac{3}{5}$
24) $\frac{3}{4}$
25) $\frac{5}{18}$
26) $\frac{25}{36}$
27) $\frac{9}{10}$
28) $\frac{8}{15}$
29) $3\frac{3}{4}$
30) $\frac{20}{33}$

Adding and Subtracting Mixed Numbers

1) $5\frac{1}{2}$
2) 8
3) $4\frac{1}{2}$
4) $4\frac{7}{12}$
5) $6\frac{5}{12}$
6) $8\frac{13}{15}$
7) $6\frac{16}{21}$
8) $7\frac{9}{10}$
9) $11\frac{29}{35}$
10) $14\frac{19}{48}$
11) $1\frac{1}{2}$
12) $2\frac{1}{5}$
13) $1\frac{2}{9}$
14) $1\frac{9}{14}$
15) $3\frac{1}{4}$
16) $3\frac{13}{15}$
17) $3\frac{1}{8}$
18) $2\frac{13}{15}$
19) $3\frac{7}{30}$
20) $4\frac{3}{35}$
21) $5\frac{1}{12}$
22) $2\frac{1}{16}$
23) $4\frac{1}{24}$
24) $2\frac{71}{72}$
25) $5\frac{33}{35}$
26) $6\frac{32}{63}$

Multiplying and Dividing Mixed Numbers

1) $12\frac{3}{8}$
2) $23\frac{1}{9}$
3) $35\frac{15}{16}$
4) $8\frac{2}{3}$
5) 5
6) $6\frac{6}{7}$
7) $30\frac{1}{3}$
8) $7\frac{6}{7}$
9) $21\frac{7}{8}$
10) 11
11) $\frac{4}{7}$
12) $1\frac{1}{4}$
13) $4\frac{2}{9}$
14) $2\frac{23}{24}$
15) 3
16) $\frac{11}{45}$
17) $\frac{19}{84}$
18) $\frac{52}{57}$
19) $4\frac{4}{27}$
20) $2\frac{26}{33}$
21) $3\frac{17}{36}$
22) $\frac{19}{22}$
23) $3\frac{27}{91}$
24) $3\frac{13}{36}$
25) $7\frac{1}{5}$
26) $4\frac{16}{45}$

Chapter 4 :

Decimals

Topics that you will practice in this chapter:

- ✓ Adding and Subtracting Decimals
- ✓ Multiplying and Dividing Decimals
- ✓ Comparing Decimals
- ✓ Rounding Decimals

"The study of mathematics, like the Nile, begins in minuteness but ends in magnificence." – Charles Caleb Colton

Adding and Subtracting Decimals

Add and subtract decimals.

1) $\begin{array}{r} 35.19 \\ -\ 24.28 \\ \hline \end{array}$

2) $\begin{array}{r} 34.29 \\ +\ 42.58 \\ \hline \end{array}$

3) $\begin{array}{r} 61.20 \\ +\ 33.75 \\ \hline \end{array}$

4) $\begin{array}{r} 38.72 \\ -\ 21.68 \\ \hline \end{array}$

5) $\begin{array}{r} 57.39 \\ +\ 26.54 \\ \hline \end{array}$

6) $\begin{array}{r} 70.24 \\ -\ 42.35 \\ \hline \end{array}$

7) $\begin{array}{r} 86.09 \\ -\ 35.14 \\ \hline \end{array}$

8) $\begin{array}{r} 54.51 \\ +\ 32.66 \\ \hline \end{array}$

9) $\begin{array}{r} 114.21 \\ -\ 88.69 \\ \hline \end{array}$

Find the missing number.

10) ___ + 2.8 = 5.4

11) 4.1 + ___ = 5.88

12) 6.45 + ___ = 8

13) 7.25 − ___ = 3.40

14) ___ − 2.35 = 4.25

15) ___ − 19.85 = 6.54

16) 22.15 + ___ = 28.95

17) ___ − 37.16 = 9.42

18) ___ + 24.50 = 34.19

19) 72.40 + ___ = 125.20

Multiplying and Dividing Decimals

Find the product.

1) $0.5 \times 0.6 =$

2) $3.3 \times 0.4 =$

3) $1.28 \times 0.5 =$

4) $0.35 \times 0.6 =$

5) $1.85 \times 0.6 =$

6) $0.24 \times 0.5 =$

7) $5.25 \times 1.4 =$

8) $18.5 \times 4.6 =$

9) $15.4 \times 6.8 =$

10) $19.5 \times 2.6 =$

11) $32.2 \times 1.5 =$

12) $78.4 \times 4.5 =$

Find the quotient.

13) $1.85 \div 10 =$

14) $74.6 \div 100 =$

15) $3.6 \div 3 =$

16) $9.6 \div 0.4 =$

17) $15.5 \div 0.5 =$

18) $32.8 \div 0.2 =$

19) $22.15 \div 1{,}000 =$

20) $53.55 \div 0.7 =$

21) $322.2 \div 0.2 =$

22) $50.67 \div 0.18 =$

23) $77.4 \div 0.8 =$

24) $27.93 \div 0.03 =$

Comparing Decimals

Write the correct comparison symbol (>, < or =).

1) 0.70 □ 0.070

2)0.049 □ 0.49

3)5.090 □ 5.09

4)2.57 □ 2.05

5)9.03 □ 0.930

6)6.06 □ 6.6

7)7.02 □ 7.020

8)3.04 □ 3.2

9)3.61 □ 3.245

10) 0.986 □ 0.0986

11) 17.24 □ 17.240

12) 0.759 □ 0.81

13) 9.040 □9.40

14) 5.73 □ 5.213

15) 9.44 □ 9.404

16) 7.17 □ 7.170

17) 4.85 □ 4.085

18) 9.041 □ 9.40

19) 3.033 □ 3.030

20) 4.97 □ 4.970

Rounding Decimals

Round each decimal to the nearest whole number.

1) 28.12
2) 6.9
3) 16.22
4) 8.5
5) 7.95
6) 52.7

Round each decimal to the nearest tenth.

7) 31.761
8) 14.421
9) 94.729
10) 77.89
11) 13.219
12) 59.89

Round each decimal to the nearest hundredth.

13) 8.428
14) 23.812
15) 55.3786
16) 231.912
17) 62.241
18) 19.447

Round each decimal to the nearest thousandth.

19) 15.54324
20) 34.62586
21) 243.8652
22) 80.4529
23) 67.1983
24) 72.36788

Convert Fraction to Decimal

Write each as a decimal.

1) $\frac{50}{100}=$

2) $\frac{46}{100}=$

3) $\frac{8}{50}=$

4) $\frac{8}{32}=$

5) $\frac{8}{72}=$

6) $\frac{56}{100}=$

7) $\frac{4}{50}=$

8) $\frac{31}{48}=$

9) $\frac{27}{300}=$

10) $\frac{15}{55}=$

11) $\frac{16}{32}=$

12) $\frac{6}{16}=$

13) $\frac{3}{10}=$

14) $\frac{18}{250}=$

15) $\frac{24}{80}=$

16) $\frac{30}{40}=$

17) $\frac{68}{100}=$

18) $\frac{7}{35}=$

19) $\frac{87}{100}=$

20) $\frac{1}{100}=$

21) $\frac{6}{36}=$

22) $\frac{2}{80}=$

Convert Decimal to Percent

Write each as a percent.

1) 0.187 =

2) 0.19 =

3) 2.6 =

4) 0.017 =

5) 0.009 =

6) 0.786 =

7) 0.245 =

8) 0.57 =

9) 0.002 =

10) 0.205 =

11) 0.324 =

12) 84.9 =

13) 3.015 =

14) 0.7 =

15) 2.35 =

16) 0.0367 =

17) 0.0043 =

18) 0.960 =

19) 6.68 =

20) 0.484 =

21) 8.957 =

22) 0.879 =

23) 2.7 =

24) 0.9 =

25) 3.6 =

26) 26.8 =

27) 1.01 =

28) 0.006 =

Convert Fraction to Percent

Write each as a percent.

1) $\frac{1}{4} =$

2) $\frac{3}{8} =$

3) $\frac{7}{14} =$

4) $\frac{15}{35} =$

5) $\frac{12}{28} =$

6) $\frac{17}{68} =$

7) $\frac{8}{11} =$

8) $\frac{14}{30} =$

9) $\frac{6}{50} =$

10) $\frac{12}{48} =$

11) $\frac{5}{34} =$

12) $\frac{27}{10} =$

13) $\frac{24}{80} =$

14) $\frac{16}{25} =$

15) $\frac{16}{58} =$

16) $\frac{2}{22} =$

17) $\frac{32}{88} =$

18) $\frac{21}{36} =$

19) $\frac{18}{92} =$

20) $\frac{6}{60} =$

21) $\frac{24}{600} =$

22) $\frac{720}{360} =$

Answers of Worksheets

Adding and Subtracting Decimals

1) 10.91
2) 76.87
3) 94.95
4) 17.04
5) 83.93
6) 27.89
7) 50.95
8) 87.17
9) 25.52
10) 2.6
11) 1.78
12) 1.55
13) 3.85
14) 6.6
15) 26.39
16) 6.8
17) 46.58
18) 9.69
19) 52.8

Multiplying and Dividing Decimals

1) 0.3
2) 1.32
3) 0.64
4) 0.21
5) 1.11
6) 0.12
7) 7.35
8) 85.1
9) 104.72
10) 50.7
11) 48.3
12) 352.8
13) 0.185
14) 0.746
15) 1.2
16) 24
17) 31
18) 164
19) 0.02215
20) 76.5
21) 1,611
22) 281.5
23) 96.75
24) 931

Comparing Decimals

1) >
2) <
3) =
4) >
5) >
6) <
7) =
8) <
9) >
10) >
11) =
12) <
13) <
14) >
15) >
16) =
17) >
18) <
19) >
20) =

Rounding Decimals

1) 28
2) 7
3) 16
4) 9
5) 8
6) 53
7) 31.8
8) 14.4
9) 94.7
10) 77.9
11) 13.2
12) 59.9
13) 8.43
14) 23.81
15) 55.38
16) 231.91
17) 62.24
18) 19.45
19) 15.543
20) 34.626
21) 243.865
22) 80.453
23) 67.198
24) 72.368

Convert Fraction to Decimal

1) 0.5
2) 0.46
3) 0.16

4) 0.25
5) 0.11
6) 0.56
7) 0.08
8) 0.646
9) 0.09
10) 0.27
11) 0.5
12) 0.375
13) 0.3
14) 0.072
15) 0.3
16) 0.75
17) 0.68
18) 0.2
19) 0.87
20) 0.01
21) 0.166
22) 0.025

Convert Decimal to Percent

1) 18.7%
2) 19%
3) 260%
4) 1.7%
5) 0.9%
6) 78.6%
7) 24.5%
8) 57%
9) 0.2%
10) 20.5%
11) 32.4%
12) 8,490%
13) 301.5%
14) 70%
15) 235%
16) 3.67%
17) 0.43%
18) 96%
19) 668%
20) 48.4%
21) 895.7%
22) 87.9%
23) 270%
24) 90%
25) 360%
26) 2,680%
27) 101%
28) 0.6%

Convert Fraction to Percent

1) 25%
2) 37.5%
3) 50%
4) 42.86%
5) 29.31%
6) 25%
7) 72.72%
8) 46.66%
9) 12%
10) 25%
11) 14.7%
12) 2.7%
13) 30%
14) 64%
15) 27.58%
16) 9.09%
17) 36.36%
18) 58.33%
19) 19.56%
20) 10%
21) 4%
22) 200%

Chapter 5 :

Proportions, Ratios, and Percent

Topics that you will practice in this chapter:

- ✓ Simplifying Ratios
- ✓ Proportional Ratios
- ✓ Similarity and Ratios
- ✓ Ratio and Rates Word Problems
- ✓ Percentage Calculations
- ✓ Percent Problems
- ✓ Discount, Tax and Tip

Without mathematics, there's nothing you can do. Everything around you is mathematics. Everything around you is numbers." – Shakuntala Devi

Simplifying Ratios

Reduce each ratio.

1) 15: 20 = ___:___

2) 7: 70 = ___:___

3) 16: 28 = ___:___

4) 7: 21 = ___:___

5) 4: 40 = ___:___

6) 6: 48 = ___:___

7) 16: 64 = ___:___

8) 10: 25 = ___:___

9) 8: 48 = ___:___

10) 49: 63 = ___:___

11) 18: 27 = ___:___

12) 35: 10 = ___:___

13) 90: 9 = ___:___

14) 24: 32 = ___:___

15) 7: 56 = ___:___

16) 45: 63 = ___:___

17) 56: 72 = ___:___

18) 26: 13 = ___:___

19) 15: 45 = ___:___

20) 28: 4 = ___:___

21) 24: 48 = ___:___

22) 30: 24 = ___:___

23) 70: 140 = ___:___

24) 6: 180 = ___:___

Write each ratio as a fraction in simplest form.

25) 6: 12 =

26) 30: 50 =

27) 15: 35 =

28) 9: 27 =

29) 8: 24 =

30) 18: 84 =

31) 7: 14 =

32) 7: 35 =

33) 40: 96 =

34) 12: 54 =

35) 44: 52 =

36) 12: 27 =

37) 15: 180 =

38) 39: 143 =

39) 20: 300 =

40) 30: 120 =

41) 56: 42 =

42) 26: 130 =

43) 66: 123 =

44) 70: 630 =

45) 75: 125 =

Proportional Ratios

Fill in the blanks; Calculate each proportion.

1) $3:8 = \_\_ : 48$
2) $2:5 = 20: \_\_$
3) $1:9 = \_\_ : 81$
4) $6:7 = 12: \_\_$
5) $9:2 = 63: \_\_$
6) $8:7 = \_\_ : 49$
7) $20:3 = \_\_ : 15$
8) $1:3 = \_\_ : 75$
9) $7:6 = \_\_ : 60$
10) $8:5 = \_\_ : 45$
11) $3:10 = 60: \_\_$
12) $6:11 = 42: \_\_$

State if each pair of ratios form a proportion.

13) $\frac{3}{20}$ *and* $\frac{9}{60}$
14) $\frac{1}{7}$ *and* $\frac{6}{42}$
15) $\frac{3}{7}$ *and* $\frac{24}{56}$
16) $\frac{4}{9}$ *and* $\frac{12}{18}$
17) $\frac{1}{9}$ *and* $\frac{12}{81}$
18) $\frac{7}{8}$ *and* $\frac{21}{28}$
19) $\frac{9}{13}$ *and* $\frac{27}{39}$
20) $\frac{1}{8}$ *and* $\frac{8}{64}$
21) $\frac{6}{19}$ *and* $\frac{30}{85}$
22) $\frac{5}{9}$ *and* $\frac{40}{81}$
23) $\frac{9}{14}$ *and* $\frac{108}{168}$
24) $\frac{15}{23}$ *and* $\frac{360}{552}$

Calculate each proportion.

25) $\frac{20}{25} = \frac{32}{x}, x = \_\_\_\_$
26) $\frac{1}{8} = \frac{32}{x}, x = \_\_\_\_$
27) $\frac{15}{5} = \frac{21}{x}, x = \_\_\_\_$
28) $\frac{1}{7} = \frac{x}{294}, x = \_\_\_\_$
29) $\frac{7}{9} = \frac{x}{81}, x = \_\_\_\_$
30) $\frac{1}{5} = \frac{13}{x}, x = \_\_\_\_$
31) $\frac{9}{5} = \frac{36}{x}, x = \_\_\_\_$
32) $\frac{6}{13} = \frac{48}{x}, x = \_\_\_\_$
33) $\frac{5}{8} = \frac{x}{88}, x = \_\_\_\_$
34) $\frac{4}{15} = \frac{x}{240}, x = \_\_\_\_$
35) $\frac{9}{19} = \frac{x}{266}, x = \_\_\_\_$
36) $\frac{7}{15} = \frac{x}{270}, x = \_\_\_\_$

Similarity and Ratios

Each pair of figures is similar. Find the missing side.

1) Right triangle: ?, 18, 12; right triangle: 5, 6, 4

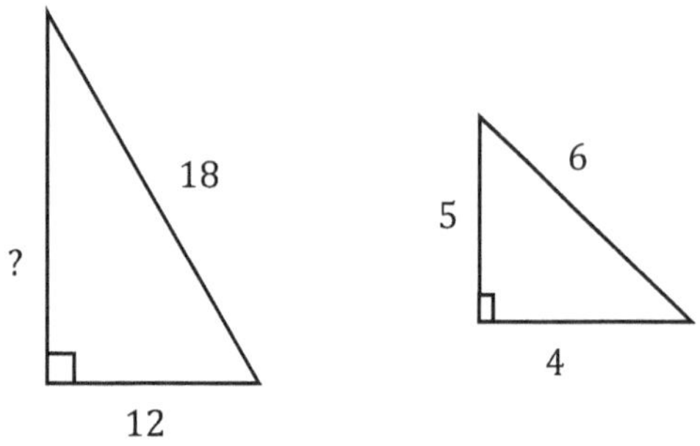

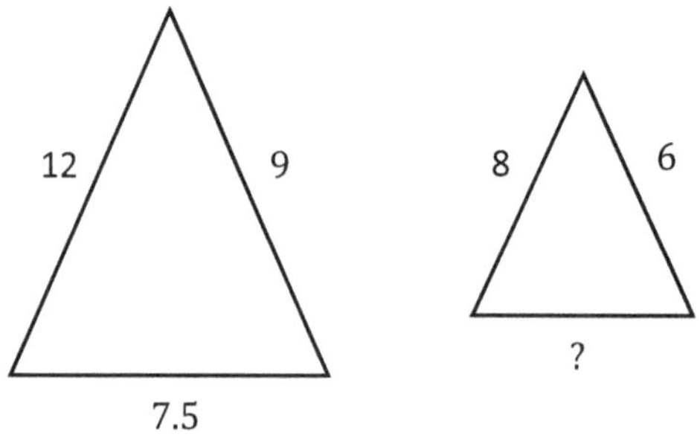

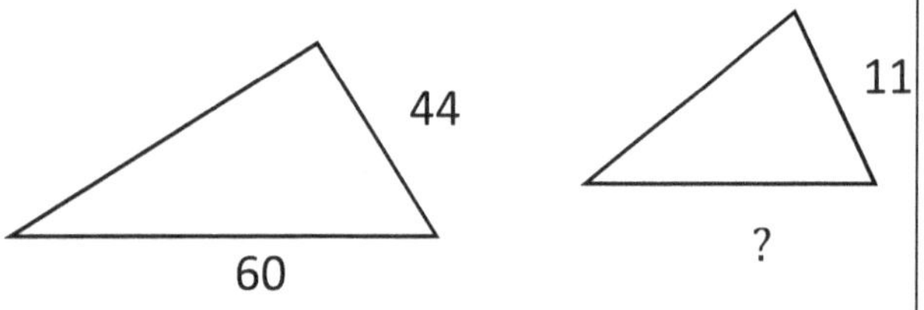

2) Triangle: 12, 9, 7.5; triangle: 8, 6, ?

3) Triangle: 44, 60; triangle: 11, ?

4) Trapezoid: ?, 7; trapezoid: 104, 56

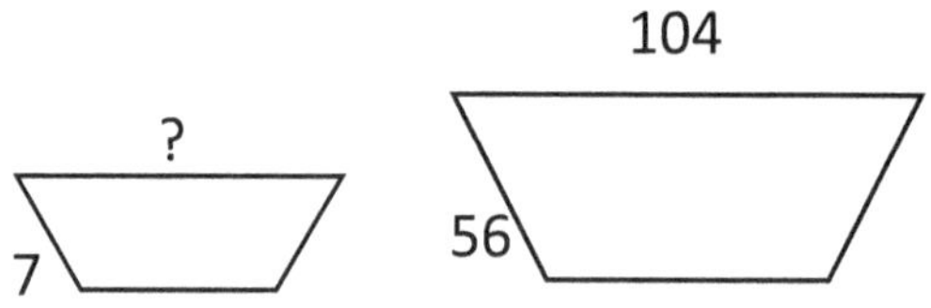

Calculate.

5) Two rectangles are similar. The first is 24 feet wide and 120 feet long. The second is 30 feet wide. What is the length of the second rectangle? ____________

6) Two rectangles are similar. One is 5 meters by 36 meters. The longer side of the second rectangle is 90 meters. What is the other side of the second rectangle? ____________

7) A building casts a shadow 25 ft long. At the same time a girl 10 ft tall casts a shadow 5 ft long. How tall is the building? ____________

8) The scale of a map of Texas is 4 inches: 32 miles. If you measure the distance from Dallas to Martin County as 38.4 inches, approximately how far is Martin County from Dallas? ________

Ratio and Rates Word Problems

Find the answer for each word problem.

1) Mason has 24 red cards and 36 green cards. What is the ratio of Mason 's red cards to his green cards? ______________

2) In a party, 45 soft drinks are required for every 54 guests. If there are 378 guests, how many soft drinks is required? ______________

3) In Mason's class, 42 of the students are tall and 24 are short. In Michael's class 84 students are tall and 48 students are short. Which class has a higher ratio of tall to short students? ______________

4) The price of 5 apples at the Quick Market is $4.6. The price of 7 of the same apples at Walmart is $5.95. Which place is the better buy? ______________

5) The bakers at a Bakery can make 90 bagels in 3 hours. How many bagels can they bake in 24 hours? What is that rate per hour? ______________

6) You can buy 5 cans of green beans at a supermarket for $5.75. How much does it cost to buy 45 cans of green beans? ______________

7) The ratio of boys to girls in a class is 4: 7. If there are 32 boys in the class, how many girls are in that class? ______________

8) The ratio of red marbles to blue marbles in a bag is 3: 7. If there are 50 marbles in the bag, how many of the marbles are red? ______________

Percentage Calculations

Calculate the given percent of each value.

1) $3\%\ of\ 60 =$ ____
2) $20\%\ of\ 32 =$ ____
3) $4\%\ of\ 72 =$ ____
4) $16\%\ of\ 32 =$ ____
5) $25\%\ of\ 124 =$ ____
6) $35\%\ of\ 56 =$ ____
7) $15\%\ of\ 20 =$ ____
8) $14\%\ of\ 150 =$ ____
9) $80\%\ of\ 50 =$ ____
10) $12\%\ of\ 115 =$ ____
11) $72\%\ of\ 250 =$ ____
12) $52\%\ of\ 500 =$ ____
13) $70\%\ of\ 400 =$ ____
14) $27\%\ of\ 145 =$ ____
15) $90\%\ of\ 64 =$ ____
16) $60\%\ of\ 55 =$ ____
17) $22\%\ of\ 210 =$ ____
18) $8\%\ of\ 235 =$ ____

Calculate the percent of each given value.

19) ____$\%\ of\ 25 = 5$
20) ____$\%\ of\ 40 = 20$
21) ____$\%\ of\ 25 = 2$
22) ____$\%\ of\ 50 = 16$
23) ____$\%\ of\ 250 = 5$
24) ____$\%\ of\ 40 = 32$
25) ____$\%\ of\ 125 = 20$
26) ____$\%\ of\ 700 = 49$
27) ____$\%\ of\ 350 = 49$
28) ____% of $500 = 210$

Calculate each percent problem.

29) A Cinema has 250 seats. 60 seats were sold for the current movie. What percent of seats are empty? _____ %

30) There are 68 boys and 92 girls in a class. 75% of the students in the class take the bus to school. How many students do not take the bus to school? _____

Percent Problems

Calculate each problem.

1) 9 is what percent of 45? ____%
2) 60 is what percent of 120? ____%
3) 10 is what percent of 200? ____%
4) 15 is what percent of 125? ____%
5) 10 is what percent of 400? ____%
6) 66 is what percent of 55? ____%
7) 40 is what percent of 160? ____%
8) 40 is what percent of 50? ____%
9) 120 is what percent of 800? ____%
10) 78 is what percent of 120? ____%
11) 36 is what percent of 144? ____%
12) 17 is what percent of 85? ____%
13) 90 is what percent of 900? ____%
14) 36 is what percent of 16? ____%
15) 63 is what percent of 14? ____%
16) 18 is what percent of 60? ____%
17) 126 is what percent of 200? ____%
18) 232 is what percent of 40? ____%

Calculate each percent word problem.

19) There are 40 employees in a company. On a certain day, 25 were present. What percent showed up for work? _____%
20) A metal bar weighs 60 ounces. 25% of the bar is gold. How many ounces of gold are in the bar? ___________
21) A crew is made up of 12 women; the rest are men. If 15% of the crew are women, how many people are in the crew? ___________
22) There are 40 students in a class and 8 of them are girls. What percent are boys? _____%
23) The Royals softball team played 400 games and won 280 of them. What percent of the games did they lose? _____%

Discount, Tax and Tip

Find the selling price of each item.

1) Original price of a computer: $420

Tax: 8%Selling price: $______

2) Original price of a laptop: $280

Tax: 4%Selling price: $______

3) Original price of a sofa: $820

Tax: 5%Selling price: $______

4) Original price of a car: $15,800

Tax: 3.6% Selling price: $______

5) Original price of a Table: $250

Tax: 9%Selling price: $______

6) Original price of a house: $630,000

Tax: 1.8% Selling price: $______

7) Original price of a tablet: $450

Discount: 30% Selling price: $____

8) Original price of a chair: $390

Discount: 8% Selling price: $____

9) Original price of a book: $75

Discount: 42% Selling price: $____

10) Original price of a cellphone: $820

Discount: 23% Selling price: $___

11) Food bill: $45

Tip: 15% Price: $______

12) Food bill: $32

Tipp: 20% Price: $______

13) Food bill: $90

Tip: 35% Price: $______

14) Food bill: $42

Tipp: 12% Price: $______

Find the answer for each word problem.

15) Nicolas hired a moving company. The company charged $500 for its services, and Nicolas gives the movers a 40% tip. How much does Nicolas tip the movers? $______

16) Mason has lunch at a restaurant and the cost of his meal is $90. Mason wants to leave a 25% tip. What is Mason's total bill including tip? $_______

17) The sales tax in Texas is 19.80% and an item costs $350. How much is the tax? $_______

18) The price of a table at Best Buy is $680. If the sales tax is 5%, what is the final price of the table including tax? $________

Answers of Worksheets

Simplifying Ratios

1) 3: 4
2) 1: 10
3) 4: 7
4) 1: 3
5) 1: 10
6) 1: 8
7) 2: 8
8) 2: 5
9) 1: 6
10) 7: 9
11) 2: 3
12) 7: 2
13) 10: 1
14) 3: 4
15) 1: 8
16) 5: 7
17) 7: 9
18) 2: 1
19) 1: 3
20) 7: 1
21) 1: 2
22) 5: 4
23) 1: 2
24) 1: 30
25) $\frac{1}{2}$
26) $\frac{3}{5}$
27) $\frac{3}{7}$
28) $\frac{1}{3}$
29) $\frac{1}{3}$
30) $\frac{3}{14}$
31) $\frac{1}{2}$
32) $\frac{1}{5}$
33) $\frac{5}{12}$
34) $\frac{2}{9}$
35) $\frac{11}{13}$
36) $\frac{4}{9}$
37) $\frac{1}{12}$
38) $\frac{3}{11}$
39) $\frac{1}{15}$
40) $\frac{1}{4}$
41) $\frac{4}{3}$
42) $\frac{1}{5}$
43) $\frac{22}{41}$
44) $\frac{1}{9}$
45) $\frac{3}{5}$

Proportional Ratios

1) 18
2) 50
3) 9
4) 14
5) 14
6) 56
7) 100
8) 25
9) 70
10) 72
11) 200
12) 77
13) Yes
14) Yes
15) Yes
16) No
17) No
18) No
19) Yes
20) Yes
21) No
22) No
23) Yes
24) Yes
25) 40
26) 256
27) 7
28) 42
29) 63
30) 65
31) 20
32) 104
33) 55
34) 64
35) 126
36) 126

Similarity and ratios

1) 15
2) 5
3) 15
4) 13
5) 150 feet
6) 12.5 meters
7) 50 feet
8) 307.2 miles

Ratio and Rates Word Problems

1) 2: 3
2) 315

3) The ratio for both classes is 7 to 4.
4) Walmart is a better buy.
5) 720, the rate is 30 per hour.
6) $51.75
7) 56
8) 15

Percentage Calculations

1) 1.8
2) 6.4
3) 2.88
4) 5.12
5) 31
6) 19.6
7) 3
8) 21
9) 40
10) 13.8
11) 180
12) 260
13) 280
14) 39.15
15) 57.6
16) 33
17) 46.2
18) 18.8
19) 20%
20) 50%
21) 8%
22) 32%
23) 2%
24) 80%
25) 16%
26) 7%
27) 14%
28) 42%
29) 76%
30) 40

Percent Problems

1) 20%
2) 50%
3) 5%
4) 12%
5) 2.5%
6) 120%
7) 25%
8) 80%
9) 15%
10) 65%
11) 25%
12) 20%
13) 10%
14) 225%
15) 450%
16) 30%
17) 63%
18) 580%
19) 62.5%
20) 15 ounces
21) 80
22) 80%
23) 30%

Discount, Tax and Tip

1) $453.60
2) $291.20
3) $861.00
4) $16,368.80
5) $272.50
6) $641,340
7) $315.00
8) $358.80
9) $43.50
10) $631.40
11) $51.75
12) $38.40
13) $121.50
14) $47.04
15) $200.00
16) $112.50
17) $69.30
18) $714.00

Chapter 6 :
Exponents and Radicals Expressions

Topics that you will practice in this chapter:

- ✓ Adding and Subtracting Exponents
- ✓ Multiplication Property of Exponents
- ✓ Zero and Negative Exponents
- ✓ Division Property of Exponents
- ✓ Powers of Products and Quotients
- ✓ Negative Exponents and Negative Bases
- ✓ Scientific Notation
- ✓ Square Roots

Mathematics is no more computation than typing is literature.

– John Allen Paulos

Adding and Subtracting Exponents

Solve each problem.

1) $3^2 + 2^5 =$

2) $x^6 + x^6 =$

3) $3b^2 - 2b^2 =$

4) $3 + 4^3 =$

5) $8 - 4^2 =$

6) $4+7^1 =$

7) $2x^3 + 3x^3 =$

8) $10^2 + 3^5 =$

9) $4^5 - 2^4 =$

10) $5^2 - 6^0 =$

11) $1^2 - 3^0 =$

12) $7^1 + 2^3 =$

13) $6^1 - 5^3 =$

14) $3^3 + 3^3 =$

15) $9^2 - 8^2 =$

16) $0^{73} + 0^{54} =$

17) $2^2 - 3^2 =$

18) $7^3 - 7^1 =$

19) $8^2 - 6^2 =$

20) $4^2 + 3^2 =$

21) $2^3 + 4^3 =$

22) $10 + 3^3 =$

23) $6x^5 + 8x^5 =$

24) $8^0 + 4^2 =$

25) $3^2 + 3^2 =$

26) $10^2 + 5^2 =$

27) $(\frac{1}{2})^2 + (\frac{1}{2})^2 =$

28) $9^2 + 3^2 =$

Multiplication Property of Exponents

Simplify and write the answer in exponential form.

1) $4 \times 4^5 =$

2) $8^4 \times 8 =$

3) $7^3 \times 7^3 =$

4) $9^2 \times 9^2 =$

5) $2^2 \times 2^4 \times 2 =$

6) $5 \times 5^3 \times 5^3 =$

7) $4^3 \times 4^2 \times 4 \times 4 =$

8) $5x \times x =$

9) $x^3 \times x^3 =$

10) $x^7 \times x^2 =$

11) $x^4 \times x^3 \times x^2 =$

12) $10x \times 3x =$

13) $4x^3 \times 4x^3 =$

14) $7x^3 \times x =$

15) $3x^2 \times 4x^2 \times x^2 =$

16) $5x^4 \times x^4 =$

17) $2x^8 \times 2x =$

18) $6x \times x^5 =$

19) $4x^2 \times 6x^6 =$

20) $5yx^3 \times 4x =$

21) $7x^3 \times y^5x^7 =$

22) $y^2x^3 \times y^5x^4 =$

23) $3x^5 \times 4x^3y^4 =$

24) $4x^4 \times 9x^2y^5 =$

25) $5x^3y^4 \times 6x^8y^2 =$

26) $8x^3y^6 \times 4xy^3 =$

27) $2xy^5 \times 6x^3y^3 =$

28) $4x^5y^2 \times 4x^2y^8 =$

29) $7x \times 3y^8x^2 \times y^5 =$

30) $x^3 \times 2y^3x^4 \times 2y =$

31) $3yx^4 \times 3y^4x \times 3xy^3 =$

32) $6y^3 \times 2y^2x^4 \times 10yx^5 =$

Zero and Negative Exponents

Evaluate the following expressions.

1) $1^{-5} =$

2) $4^{-1} =$

3) $0^{10} =$

4) $1^{15} =$

5) $5^{-2} =$

6) $3^{-3} =$

7) $9^{-1} =$

8) $10^{-2} =$

9) $12^{-2} =$

10) $2^{-5} =$

11) $3^{-4} =$

12) $2^{-4} =$

13) $6^{-3} =$

14) $10^{-3} =$

15) $30^{-1=}$

16) $15^{-2} =$

17) $4^{-3} =$

18) $2^{-7} =$

19) $5^{-3} =$

20) $4^{-4} =$

21) $3^{-5} =$

22) $10^{-4} =$

23) $2^{-10} =$

24) $8^{-3} =$

25) $20^{-2} =$

26) $14^{-2} =$

27) $9^{-3} =$

28) $100^{-2} =$

29) $5^{-4} =$

30) $4^{-6} =$

31) $(\frac{1}{4})^{-3}$

32) $(\frac{1}{6})^{-2} =$

33) $(\frac{1}{7})^{-2} =$

34) $(\frac{2}{3})^{-3} =$

35) $(\frac{1}{13})^{-2} =$

36) $(\frac{7}{12})^{-2} =$

37) $(\frac{1}{6})^{-3} =$

38) $(\frac{1}{300})^{-2} =$

39) $(\frac{2}{9})^{-2} =$

40) $(\frac{7}{5})^{-1} =$

41) $(\frac{13}{23})^{0} =$

42) $(\frac{1}{4})^{-5} =$

Division Property of Exponents

Simplify.

1) $\frac{5^6}{5^7} =$

2) $\frac{8^8}{8^6} =$

3) $\frac{4^5}{4} =$

4) $\frac{3}{3^5} =$

5) $\frac{x}{x^6} =$

6) $\frac{3\times 3^2}{3^2\times 3^5} =$

7) $\frac{9^4}{9^2} =$

8) $\frac{10\times 10^9}{10^2\times 10^7} =$

9) $\frac{7^5\times 7^7}{7^4\times 7^8} =$

10) $\frac{15x}{30x^6} =$

11) $\frac{3x^9}{4x^4} =$

12) $\frac{15x^8}{10x^9} =$

13) $\frac{42x^5}{6y^9} =$

14) $\frac{36y^8}{4x^4y^5} =$

15) $\frac{2x^7}{9x} =$

16) $\frac{49x^8y^6}{7x^9} =$

17) $\frac{48x^2}{24x^6y^{12}} =$

18) $\frac{30yx^5}{6yx^7} =$

19) $\frac{19x^7y}{38x^{12}y^4} =$

20) $\frac{9x^8}{63x^8} =$

21) $\frac{9x^{-9}}{4x^{-3}} =$

Powers of Products and Quotients

Simplify.

1) $(4^3)^2 =$

2) $(2^3)^4 =$

3) $(2 \times 2^3)^2 =$

4) $(5 \times 5^5)^6 =$

5) $(19^4 \times 19^2)^3 =$

6) $(2^3 \times 2^4)^4 =$

7) $(5 \times 5^2)^2 =$

8) $(4^4)^4 =$

9) $(8x^5)^2 =$

10) $(3x^2y^4)^4 =$

11) $(7x^5y^2)^2 =$

12) $(5x^4y^4)^3 =$

13) $(2x^3y^3)^5 =$

14) $(10x^3y^4)^3 =$

15) $(13y^3y)^2 =$

16) $(5x^6x^4)^2 =$

17) $(6x^7y^6)^3 =$

18) $(12x^5x^7)^2 =$

19) $(2x^4 \times 2x)^4 =$

20) $(2x^4y^3)^5 =$

21) $(15x^7y^2)^2 =$

22) $(8x^3y^5)^3 =$

23) $(3x \times 2y^2)^4 =$

24) $(\frac{4x}{x^5})^2 =$

25) $\left(\frac{x^4y^5}{x^3y^5}\right)^9 =$

26) $\left(\frac{36xy}{6x^5}\right)^3 =$

27) $\left(\frac{x^7}{x^8y^2}\right)^6 =$

28) $\left(\frac{xy^4}{x^3y^6}\right)^{-3} =$

29) $\left(\frac{5xy^8}{x^3}\right)^2 =$

30) $\left(\frac{xy^6}{2xy^3}\right)^{-4} =$

Negative Exponents and Negative Bases

Simplify.

1) $-9^{-1} =$

2) $-9^{-2} =$

3) $-2^{-5} =$

4) $-x^{-7} =$

5) $11x^{-1} =$

6) $-8x^{-3} =$

7) $-12x^{-5} =$

8) $-9x^{-8}y^{-6} =$

9) $32x^{-5}y^{-1} =$

10) $10a^{-9}b^{-3} =$

11) $-17x^{4}y^{-6} =$

12) $-\frac{25}{x^{-5}} =$

13) $-\frac{13x}{a^{-7}} =$

14) $(-\frac{1}{3})^{-4} =$

15) $(-\frac{3}{4})^{-2} =$

16) $-\frac{14}{a^{-6}b^{-3}} =$

17) $-\frac{7x}{x^{-8}} =$

18) $-\frac{a^{-9}}{b^{-5}} =$

19) $-\frac{11}{x^{-5}} =$

20) $\frac{8b}{-16c^{-6}} =$

21) $\frac{12ab}{a^{-4}b^{-3}} =$

22) $-\frac{8n^{-4}}{32p^{-7}} =$

23) $\frac{16ab^{-6}}{-6c^{-5}} =$

24) $(\frac{10a}{5c})^{-4} =$

25) $(-\frac{12x}{4yz})^{-3} =$

26) $\frac{8ab^{-7}}{-5c^{-3}} =$

27) $(-\frac{x^{4}}{x^{5}})^{-5} =$

28) $(-\frac{x^{-2}}{7x^{3}})^{-2} =$

29) $(-\frac{x^{-4}}{x^{2}})^{-6} =$

Scientific Notation

Write each number in scientific notation.

1) 0.223 =

2) 0.09 =

3) 4.5 =

4) 900 =

5) 2,000 =

6) 0.006 =

7) 33 =

8) 9,400 =

9) 1,470 =

10) 52,000 =

11) 8,000,000 =

12) 0.00009 =

13) 2,158,000 =

14) 0.0039 =

15) 0.000075 =

16) 4,300,000 =

17) 130,000 =

18) 4,000,000,000 =

19) 0.00009 =

20) 0.0039 =

Write each number in standard notation.

21) $4 \times 10^{-1} =$

22) $1.2 \times 10^{-3} =$

23) $2.7 \times 10^{5} =$

24) $6 \times 10^{-4} =$

25) $3.6 \times 10^{-3} =$

26) $5.5 \times 10^{5} =$

27) $3.2 \times 10^{4} =$

28) $3.88 \times 10^{6} =$

29) $7 \times 10^{-6} =$

30) $4.2 \times 10^{-7} =$

Square Roots

Find the value each square root.

1) $\sqrt{16} =$ ____
2) $\sqrt{25} =$ ____
3) $\sqrt{1} =$ ____
4) $\sqrt{64} =$ ____
5) $\sqrt{0} =$ ____
6) $\sqrt{196} =$ ____
7) $\sqrt{4} =$ ____
8) $\sqrt{256} =$ ____
9) $\sqrt{36} =$ ____
10) $\sqrt{289} =$ ____
11) $\sqrt{169} =$ ____
12) $\sqrt{144} =$ ____
13) $\sqrt{100} =$ ____
14) $\sqrt{1{,}600} =$ ____
15) $\sqrt{2{,}500} =$ ____
16) $\sqrt{324} =$ ____
17) $\sqrt{529} =$ ____
18) $\sqrt{20} =$ ____
19) $\sqrt{625} =$ ____
20) $\sqrt{18} =$ ____
21) $\sqrt{50} =$ ____
22) $\sqrt{1{,}024} =$ ____
23) $\sqrt{160} =$ ____
24) $\sqrt{32} =$ ____

Evaluate.

25) $\sqrt{4} \times \sqrt{25} =$ ________
26) $\sqrt{36} \times \sqrt{49} =$ ________
27) $\sqrt{6} \times \sqrt{6} =$ ________
28) $\sqrt{13} \times \sqrt{13} =$ ________
29) $2\sqrt{5} \times 3\sqrt{5} =$ ________
30) $\sqrt{12} \times \sqrt{3} =$ ________
31) $\sqrt{13} + \sqrt{13} =$ ________
32) $\sqrt{10} + 2\sqrt{10} =$ ________
33) $12\sqrt{7} - 10\sqrt{7} =$ ________
34) $4\sqrt{10} \times 2\sqrt{10} =$ ________
35) $5\sqrt{3} \times 8\sqrt{3} =$ ________
36) $6\sqrt{3} - \sqrt{12} =$ ________

Answers of Worksheets

Add and Subtract Exponents.

1) 41
2) $2x^6$
3) b^2
4) 67
5) -8
6) 11
7) $5x^3$
8) 343
9) 1,008
10) 24
11) 0
12) 15
13) -119
14) 54
15) 17
16) 0
17) -5
18) 336
19) 28
20) 25
21) 72
22) 37
23) $14x^5$
24) 17
25) 18
26) 125
27) $\frac{1}{2}$
28) 90

Multiplication Property of Exponents

1) 4^6
2) 8^5
3) 7^6
4) 9^4
5) 2^7
6) 5^7
7) 4^7
8) $5x^2$
9) x^6
10) x^9
11) x^9
12) $30x^2$
13) $16x^6$
14) $7x^4$
15) $12x^6$
16) $5x^8$
17) $4x^9$
18) $6x^6$
19) $24x^8$
20) $20x^4y$
21) $7x^{10}y^5$
22) x^7y^7
23) $12x^8y^4$
24) $36x^6y^5$
25) $30x^{11}y^6$
26) $32x^4y^9$
27) $12x^4y^8$
28) $16x^7y^{10}$
29) $21x^3y^{13}$
30) $4x^7y^4$
31) $27x^6y^8$
32) $120x^9y^6$

Zero and Negative Exponents

1) 1
2) $\frac{1}{4}$
3) 0
4) 1
5) $\frac{1}{25}$
6) $\frac{1}{27}$
7) $\frac{1}{9}$
8) $\frac{1}{100}$
9) $\frac{1}{144}$
10) $\frac{1}{32}$
11) $\frac{1}{81}$
12) $\frac{1}{16}$
13) $\frac{1}{216}$
14) $\frac{1}{1,000}$
15) $\frac{1}{30}$
16) $\frac{1}{225}$
17) $\frac{1}{64}$
18) $\frac{1}{128}$
19) $\frac{1}{125}$
20) $\frac{1}{256}$
21) $\frac{1}{243}$
22) $\frac{1}{10,000}$
23) $\frac{1}{1,024}$
24) $\frac{1}{512}$
25) $\frac{1}{400}$

26) $\frac{1}{196}$
27) $\frac{1}{729}$
28) $\frac{1}{10,000}$
29) $\frac{1}{625}$
30) $\frac{1}{4,096}$
31) 64
32) 36
33) 49
34) $\frac{27}{8}$
35) 169
36) $\frac{144}{49}$
37) 216
38) 90,000
39) $\frac{81}{4}$
40) $\frac{5}{7}$
41) 1
42) 1,024

Division Property of Exponents

1) $\frac{1}{5}$
2) 8^2
3) 4^4
4) $\frac{1}{3^4}$
5) $\frac{1}{x^5}$
6) $\frac{1}{3^4}$
7) 9^2
8) 10
9) 1
10) $\frac{1}{2x^5}$
11) $\frac{3x^5}{4}$
12) $\frac{3}{2x}$
13) $\frac{7x^5}{y^9}$
14) $\frac{9y^3}{x^4}$
15) $\frac{2x^6}{9}$
16) $\frac{7y^6}{x}$
17) $\frac{2}{x^4y^{12}}$
18) $\frac{5}{x^2}$
19) $\frac{1}{2x^5y^3}$
20) $\frac{1}{7}$
21) $\frac{9}{4x^6}$

Powers of Products and Quotients

1) 4^6
2) 2^{12}
3) 2^8
4) 5^{36}
5) 19^{18}
6) 2^{28}
7) 5^6
8) 4^{16}
9) $64x^{10}$
10) $81x^8y^{16}$
11) $49x^{10}y^4$
12) $125x^{12}y^{12}$
13) $32x^{15}y^{15}$
14) $1,000x^9y^{12}$
15) $169y^8$
16) $25x^{20}$
17) $216x^{21}y^{18}$
18) $144x^{24}$
19) $256x^{20}$
20) $32x^{20}y^{15}$
21) $225x^{14}y^4$
22) $512x^9y^{15}$
23) $1,296x^4y^8$
24) $\frac{16}{x^8}$
25) x^9
26) $\frac{216y^3}{x^{12}}$
27) $\frac{1}{x^6y^{12}}$
28) x^6y^6
29) $\frac{25y^{16}}{x^4}$
30) $\frac{16}{y^{12}}$

Negative Exponents and Negative Bases

1) $-\frac{1}{9}$
2) $-\frac{1}{81}$
3) $-\frac{1}{32}$
4) $-\frac{1}{x^7}$
5) $\frac{11}{x}$
6) $-\frac{8}{x^3}$
7) $-\frac{12}{x^5}$
8) $-\frac{9}{x^8y^6}$
9) $\frac{32}{x^5y}$

10) $\frac{10}{a^9b^3}$
11) $-\frac{17x^4}{y^6}$
12) $-25x^5$
13) $-13xa^7$
14) 81
15) $\frac{16}{9}$
16) $-14a^6b^3$
17) $-7x^9$
18) $-\frac{b^5}{a^9}$
19) $-11x^5$
20) $-\frac{bc^6}{2}$
21) $12a^5b^4$
22) $-\frac{p^7}{4n^4}$
23) $-\frac{8ac^5}{3b^6}$
24) $\frac{c^4}{16a^4}$
25) $\frac{y^3z^3}{27x^3}$
26) $-\frac{8ac^3}{5b^7}$
27) $-x^5$
28) $49x^{10}$
29) x^{36}

Scientific Notation

1) 2.23×10^{-1}
2) 9×10^{-2}
3) 4.5×10^{0}
4) 9×10^{2}
5) 2×10^{3}
6) 6×10^{-3}
7) 3.3×10^{1}
8) 9.4×10^{3}
9) 1.47×10^{3}
10) 5.2×10^{4}
11) 8×10^{6}
12) 9×10^{-5}
13) 2.158×10^{6}
14) 3.9×10^{-3}
15) 7.5×10^{-5}
16) 4.3×10^{6}
17) 1.3×10^{5}
18) 4×10^{9}
19) 9×10^{-5}
20) 3.9×10^{-3}
21) 0.4
22) 0.0012
23) 270,000
24) 0.0006
25) 0.0036
26) 550,000
27) 32,000
28) 3,880,000
29) 0.000007
30) 0.00000042

Square Roots

1) 4
2) 5
3) 1
4) 8
5) 0
6) 14
7) 2
8) 16
9) 6
10) 17
11) 13
12) 12
13) 10
14) 40
15) 50
16) 18
17) 23
18) $2\sqrt{5}$
19) 25
20) $3\sqrt{2}$
21) $5\sqrt{2}$
22) 32
23) $4\sqrt{10}$
24) $4\sqrt{2}$
25) 10
26) 42
27) 6
28) 13
29) 30
30) 6
31) $2\sqrt{13}$
32) $3\sqrt{10}$
33) $2\sqrt{7}$
34) 80
35) 120
36) $4\sqrt{3}$

Chapter 7 : Measurements

Topics that you will learn in this chapter:

- ✓ Reference Measurement
- ✓ Metric Length
- ✓ Customary Length
- ✓ Metric Capacity
- ✓ Customary Capacity
- ✓ Metric Weight and Mass
- ✓ Customary Weight and Mass
- ✓ Temperature
- ✓ Time

"It's not that I'm so smart, it's just that I stay with problems longer." -Albert Einstein

Reference Measurement

LENGTH	
Customary	**Metric**
1 mile (mi) = 1,760 yards (yd)	1 kilometer (km) = 1,000 meters (m)
1 yard (yd) = 3 feet (ft)	1 meter (m) = 100 centimeters (cm)
1 foot (ft) = 12 inches (in.)	1 centimeter(cm) = 10 millimeters(mm)
VOLUME AND CAPACITY	
Customary	**Metric**
1 gallon (gal) = 4 quarts (qt)	1 liter (L) = 1,000 milliliters (mL)
1 quart (qt) = 2 pints (pt.)	
1 pint (pt.) = 2 cups (c)	
1 cup (c) = 8 fluid ounces (Fl oz)	
WEIGHT AND MASS	
Customary	**Metric**
1 ton (T) = 2,000 pounds (lb.)	1 kilogram (kg) = 1,000 grams (g)
1 pound (lb.) = 16 ounces (oz)	1 gram (g) = 1,000 milligrams (mg)

Time
1 year = 12 months
1 year = 52 weeks
1 week = 7 days
1 day = 24 hours
1 hour = 60 minutes
1 minute = 60 seconds

Metric Length Measurement

Convert to the units.

1) 3×10^3 mm = _______ cm
2) 0.95 m = __________ mm
3) 0.08 m = __________ cm
4) 2.25 km = __________ m
5) 7,800 mm =______ m
6) 9,100 cm =________ m
7) 5.83 m = __________ cm
8) 2×10^5mm =______ cm
9) 8×10^3 mm = ______ m
10) 0.003 km =_________ mm
11) 0.7 km = ________ m
12) 0.011 m = _________ cm
13) 125×10^5m = _______ km
14) 78×10^4 m = _______ km

Customary Length Measurement

Convert to the units.

1) 15 ft = _____ in
2) 1.5 ft = _____ in
3) 4.8 yd = _____ ft
4) 0.82 yd = _____ ft
5) 17×10^{-3} yd = _____ in
6) 0.5 mi = _____ in
7) 1,746 in =____ yd
8) 3.24 in =______ yd
9) 3,960 yd = ______ mi
10) 42.55 yd =______ in
11) 5×10^{-2}mi = ______ yd
12) 87,120ft =______ mi
13) 2.52 in = ______ ft
14) 29.3 yd = ______ feet
15) 0.612 in = _____ ft
16) 1.3 mi = _____ ft

Metric Capacity Measurement

Convert the following measurements.

1) 1.58 l = __________ ml
2) 0.504 l = __________ ml
3) 3.04 l = __________ ml
4) 0.005 l = __________ ml
5) 121.56 l = __________ ml
6) 0.0459 l = __________ ml
7) 4.2×10^5 ml = __________ l
8) 3.12×10^3 ml = ______ l
9) $1{,}889 \times 10^2$ ml = ________ l
10) 250ml = ________ l
11) 656,160 ml = ________ l
12) 0.54×10^4 ml = _______ l

Customary Capacity Measurement

Convert the following measurements.

1) 0.7 gal = ______ qt.
2) 3.2 gal = ______ pt.
3) 0.75 gal = _______ c.
4) 15.5 pt. = _______ c
5) 18.2 c = _______ fl oz
6) 9.02 qt = ________ pt.
7) 1.05 qt = ________ c
8) 158 pt. = _______ c
9) 9.6×10^3 c = _________ gal
10) 203.2 pt. = ______ gal
11) 12.4 qt = ______ gal
12) 115.6 pt. = ______ qt
13) 4,880 c = ________ qt
14) 113.6 c = _______ pt.
15) 0.036 qt= _______ gal
16) 522.4 pt. = ______ qt
17) 5.8 gal = _______ pt.
18) 0.002 qt = ________ c
19) 672 c = _________ gal
20) 72.96 fl oz = _______ c

Metric Weight and Mass Measurement

Convert.

1) 0.712 kg = __________ g
2) 54.01 kg = __________ g
3) 9.8×10^{-5} kg = __________ g
4) 0.012 kg = __________ g
5) 120.02 kg = __________ g
6) 1.199 kg = __________ g
7) 0.0055 kg = __________ g
8) 9×10^{4} g = __________ kg
9) 3.5×10^{5} g = __________ kg
10) 0.008×10^{4} g = __________ kg
11) 15,010 g = __________ kg
12) 12.1×10^{4} g = __________ kg
13) 4,155,200 g = __________ kg
14) 402×10^{2} g = __________ kg

Customary Weight and Mass Measurement

Convert.

1) 36×10^{2} lb. = ______ T
2) 0.022×10^{4} lb. = _____ T
3) 215,000 lb. = ______ T
4) 12,600 lb. = _____ T
5) 0.015 lb. = _________ oz
6) 1.6 lb. = _________ oz
7) 0.021 lb. = ________ oz
8) 5.2 T = ______ lb.
9) 6.8×10^{-3} T = ______ lb.
10) 156×10^{-2} T = ______ lb.
11) 0.017 T = ______ lb.
12) 1.085 T = _______ oz
13) 0.006 T = _______ oz
14) 209.92 oz = _______ lb.

Temperature

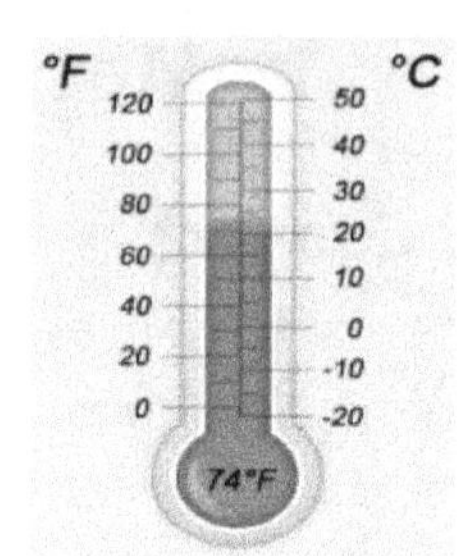

Convert Fahrenheit into Celsius.

1) 35.6°F = ___ °C
2) 54.5°F= ___ °C
3) −25.6°F= ___ °C
4) 62.6°F= ___ °C
5) 120.2°F= ___ °C
6) 174.2°F= ___ °C
7) 51.8°F= ___ °C
8) 193.1°F= ___ °C
9) 221°F= ___ °C
10) 60.44°F= ___ °C
11) 48.2°F= ___ °C
12) 134.6°F= ___ °C

Convert Celsius into Fahrenheit.

13) 18.2°C = __ °F
14) 88.8°C = __ °F
15) 250°C = __°F
16) 52°C = __ °F
17) 10°C = __ °F
18) −24°C = __ °F
19) 6°C = __ °F
20) 15.6°C = __°F
21) 33°C = __ °F
22) 61°C = __ °F
23) 113.5°C = __°F
24) 28°C = __°F

Time

Convert to the units.

1) 18.5 hr. = __________ min
2) 26 year = ________ week
3) 0.2 hr. = ___________ sec
4) 12.5 min = _________ sec
5) 1.2×10^5 min = _______ hr
6) 1,460 day = ________ year
7) 3 year = ___________ hr.
8) 51 day = _________ hr
9) 5 day = _________ min
10) 552 min = ________ hr
11) 32.25 year = ______ month
12) 6,480 sec = _______ min
13) 264 hr = _________ day
14) 22 weeks = _______ day

How much time has passed?

1) From 1:45 A.M. to 4:55 A.M.: ____ hours and ___ minutes.
2) From 1:25 A.M. to 6:05 A.M.: ____ hours and ___ minutes.
3) It's 7:15 P.M. What time was 4 hours ago? _______ O'clock
4) 3:05 A.M to 6:55 AM: _____ hours and _____ minutes.
5) 3:45 A.M to 5:15 AM: _____ hours and _____ minutes.
6) 8:05 A.M. to 11:20 AM. = _____ hour(s) and _____ minutes.
7) 10:55 A.M. to 1:25 PM. = _____ hour(s) and _____ minutes
8) 6:18 A.M. to 6:52 A.M. = _______ minutes
9) 3:54 A.M. to 4:08 A.M. = _______ minutes

Answers of Worksheets

Metric length

1) 300 cm
2) 950 mm
3) 8 cm
4) 2,250 m
5) 7.8 m
6) 91 m
7) 583 cm
8) 20,000 cm
9) 8 m
10) 3,000 mm
11) 700 m
12) 1.1 cm
13) 12,500 km
14) 780km

Customary Length

1) 180
2) 18
3) 14.4
4) 2.46
5) 0.612
6) 31,680
7) 48.5
8) 0.09
9) 2.25
10) 1,531.8
11) 88
12) 16.5
13) 0.21
14) 87.9
15) 0.051
16) 6,864

Metric Capacity

1) 1,580 ml
2) 504 ml
3) 3,040 ml
4) 5 ml
5) 121,560 ml
6) 45.9 ml
7) 420 L
8) 3.12 L
9) 188.9 L
10) 0.25L
11) 656.16 L
12) 5.4 L

Customary Capacity

1) 2.8 qt
2) 25.6 pt.
3) 12 c
4) 31 c
5) 145.6 fl oz
6) 18.04 pt.
7) 4.2 c
8) 316 c
9) 600 gal
10) 25.4 gal
11) 3.1 gal
12) 57.8 qt
13) 1,220qt
14) 56.8 pt.
15) 0.009 gal
16) 261.2 qt
17) 46.4 pt.
18) 0.008 c
19) 42 gal
20) 9.12 c

Metric Weight and Mass

1) 712 g
2) 54,010 g
3) 0.098 g
4) 12 g
5) 120,020 g
6) 1,199 g
7) 5.5 g
8) 90 kg
9) 350 kg
10) 0.08 kg
11) 15.01 kg
12) 121 kg
13) 4,155.2 kg
14) 40.2 kg

Customary Weight and Mass

1) 1.8 T
2) 0.11 T
3) 107.5 T
4) 6.3 T
5) 0.24 oz
6) 25.6 oz
7) 0.336 oz
8) 10,400 lb.
9) 13.6 lb.
10) 3,120 lb.
11) 34 lb.
12) 34,720 oz
13) 192 oz
14) 13.12 lb

Temperature

1) 2°C
2) 12.5°C
3) −32°C
4) 17°C
5) 49°C
6) 79°C
7) 11°C
8) 89.5°C
9) 105°C
10) 15.8°C
11) 9°C
12) 57°C
13) 64.76°F
14) 191.84°F
15) 482°F
16) 125.6°F
17) 50°F
18) −11.2°F
19) 42.8°F
20) 60.08°F
21) 91.4°F
22) 141.8°F
23) 236.3°F
24) 82.4°F

Time - Convert.

1) 1,110 min
2) 1,352 weeks
3) 720 sec
4) 750 sec
5) 2,000 hr
6) 4 year
7) 26,280 hr
8) 1,224 hr
9) 7,200 min
10) 9.2 hr
11) 387 months
12) 108 min
13) 11 days
14) 154 days

Time - Gap

1) 3:10
2) 4:40
3) 3:15 P.M.
4) 3:50
5) 1:30
6) 3:15
7) 2:30
8) 34 minutes
9) 14 minutes

Chapter 8 :
Algebraic Expressions

Topics that you will practice in this chapter:

- ✓ Find a rule!
- ✓ Translate Phrases into an Algebraic Statement
- ✓ Simplifying Variable Expressions
- ✓ The Distributive Property
- ✓ Evaluating One Variable Expressions
- ✓ Evaluating Two Variables Expressions
- ✓ Combining like Terms

Mathematics is, as it were, a sensuous logic, and relates to philosophy as do the arts, music, and plastic art to poetry. — K. Shegel

Find a Rule!

Complete the output.

1- **Rule:** the output is $x - 10.5$

Input	x	15	18	27	32.25	48.5
Output	y					

1) **Rule:** the output is $x \times 5\frac{1}{3}$

Input	x	3	9	15	21	33
Output	y					

2- **Rule:** the output is $x \div 9$

Input	x	513	387	342	198	126
Output	y					

Find a rule to write an expression.

3- **Rule:** ____________________

Input	x	4	14	19	24
Output	y	10	35	47.5	60

4- **Rule:** ____________________

Input	x	5	13	19.6	34.5
Output	y	14.4	22.4	29	43.9

5- **Rule:** ____________________

Input	x	72	96	132	230.4
Output	y	9	12	16.5	28.8

Translate Phrases into an Algebraic Statement

Write an algebraic expression for each phrase.

1) 9 multiplied by x. ________________

2) Subtract 11 from y. ________________

3) 19 divided by x. ________________

4) 38 decreased by y. ________________

5) Add y to 40. ________________

6) The square of 6. ________________

7) x raised to the fifth power. ________________

8) The sum of six and a number. ________________

9) The difference between fifty–seven and y. ________________

10) The quotient of nine and a number. ________________

11) The quotient of the square of x and 25. ________________

12) The difference between x and 6 is 19. ________________

13) 10 times a reduced by the square of b. ________________

14) Subtract the product of a and b from 41. ________________

Simplifying Variable Expressions

Simplify each expression.

1) $3(x+5)=$

2) $(-4)(7x-5)=$

3) $11x+5-6x=$

4) $-4-2x^2-6x^2=$

5) $7+13x^2+3=$

6) $3x^2+7x+15x^2=$

7) $3x^2-12x^2+4x=$

8) $4x^2-8x-2x=$

9) $6x+7(3-4x)=$

10) $8x+4(15x-3)=$

11) $6(-3x-9)-17=$

12) $-11x^2-(-5x)=$

13) $2x+7+5-8x=$

14) $7+6x-11-5x=$

15) $27x+8-13-5x=$

16) $(-11)(-5x+2)-41x=$

17) $19x-4(4-2x)=$

18) $16x+3(3x+6)+10=$

19) $5(-2x-4)-13x=$

20) $16x-3x(x+10)=$

21) $17x+5x(2-4x)=$

22) $5x(-4x-7)+20x=$

23) $25x-19+4x^2=$

24) $6x(x-11)+25=$

25) $4x-5+15x+3x^2=$

26) $-7x^2-11x-9x=$

27) $10x-9x^2-3x^2-7=$

28) $13+3x^2-9x^2-21x=$

29) $22x+10x^2-15x+17=$

30) $4x^2+25x+21x^2=$

31) $29-12x^2-23x-4x^2=$

32) $22x-19x-9x^2+30=$

The Distributive Property

Use the distributive property to simply each expression.

1) $4(1 + 2x) =$

2) $2(4 + 7x) =$

3) $3(4x - 4) =$

4) $(2x - 5)(-6) =$

5) $(-3)(x + 6) =$

6) $(4 + 3x)2 =$

7) $(-5)(8 - 3x) =$

8) $-(-5 - 7x) =$

9) $(-6x + 3)(-3) =$

10) $(-4)(x - 7) =$

11) $-(5 - 3x) =$

12) $3(9 + 4x) =$

13) $6(4 + 3x) =$

14) $(-5x + 3)2 =$

15) $(5 - 8x)(-3) =$

16) $(-12)(3x + 3) =$

17) $(5 - 3x)6 =$

18) $4(2 + 6x) =$

19) $8(7x - 3) =$

20) $(-2x + 3)4 =$

21) $(7 - 5x)(-9) =$

22) $(-10)(x - 8) =$

23) $(11 - 4x)3 =$

24) $(-6)(10x - 4) =$

25) $(3 - 9x)(-7) =$

26) $(-9)(x + 9) =$

27) $(-3 + 5x)(-7) =$

28) $(-5)(8 - 10x) =$

29) $12(4x - 8) =$

30) $(-10x + 13)(-3) =$

31) $(-8)(3x - 2) + 4(x + 5) =$

32) $(-8)(x + 4) - (6 + 5x) =$

Evaluating One Variable Expressions

Evaluate each expression using the value given.

1) $8 - x, x = 5$

2) $x - 9, x = 5$

3) $5x + 4, x = 3$

4) $x - 13, x = -4$

5) $12 - x, x = 4$

6) $x + 2, x = 6$

7) $4x + 8, x = 3$

8) $x + (-7), x = -8$

9) $4x + 5, x = 2$

10) $3x + 9, x = -2$

11) $15 + 3x - 7, x = 2$

12) $17 - 3x, x = 3$

13) $8x - 9, x = 4$

14) $5x + 4, x = -3$

15) $10x + 5, x = 3$

16) $14 - 4x, x = -6$

17) $3(5x + 3), x = 9$

18) $4(-3x - 6), x = 3$

19) $7x - 2x + 12, x = 4$

20) $(5x + 6) \div 2, x = 8$

21) $(x + 18) \div 10, x = 12$

22) $5x - 12 + 3x, x = -3$

23) $(6 - 4x)(-3), x = -4$

24) $9x^2 + 3x - 6, x = 2$

25) $x^2 - 10x, x = -5$

26) $3x(7 - 2x), x = 2$

27) $12x + 6 - 2x^2, x = -4$

28) $(-3)(4x - 8 + 3x), x = 3$

29) $(-6) + \frac{x}{4} + 3x, x = 16$

30) $(-6) + \frac{x}{5}, x = 35$

31) $\left(-\frac{45}{x}\right) - 7 + 2x, x = 9$

32) $\left(-\frac{21}{x}\right) - 12 + 4x, x = 7$

Combining like Terms

Simplify each expression.

1) $11x + 3x + 6 =$

2) $8(2x - 6) =$

3) $18x - 7x + 11 =$

4) $(-4)(6x - 7) =$

5) $22x - 10x - 5 =$

6) $32x - 13 + 8x =$

7) $15 - (8x - 11) =$

8) $-24x + 17 - 11x =$

9) $12x - 8 - 6x + 9 =$

10) $21x + 5 - 36 + 12x =$

11) $28x + 3x - 11 =$

12) $(-3x + 4)5 =$

13) $2 + 4x + 9x - 8 =$

14) $6(2x - 5x) - 4 =$

15) $4(5x + 11) + 3x =$

16) $x - 14 - 11x =$

17) $5(10 + 9x) - 8x =$

18) $42x + 17 - 23x =$

19) $(-7x) + 19 + 20x =$

20) $(-7x) - 33 + 29x =$

21) $4(5x + 3) - 19x =$

22) $5(6 - 2x) - 15x =$

23) $-24x + (11 - 18x) =$

24) $(-9) - (6)(7x + 3) =$

25) $(-1)(8x - 10) - 21x =$

26) $-36x + 14 + 27x - 5x =$

27) $3(-13x + 6) - 17x =$

28) $-5x - 42 + 32x =$

29) $37x - 19x + 15 - 9x =$

30) $3(5x + 7x) - 31 =$

31) $14 - 6x - 15 - 9x =$

32) $-2(-5x - 7x) + 27x =$

Answers of Worksheets

Find a rule.

1)

Input	x	15	18	27	32.25	48.5
Output	y	4.5	7.5	16.5	21.75	38

2)

Input	x	3	9	15	21	33
Output	y	16	48	80	112	176

3)

Input	x	513	387	342	198	126
Output	y	57	43	38	22	14

4) y = 2.5x

5) y = x + 9.4

6) y = x ÷ 8

Translate Phrases into an Algebraic Statement

1) $9x$
2) $y - 11$
3) $\frac{19}{x}$
4) $38 - y$
5) $y + 40$
6) 6^2
7) x^5
8) $6 + x$
9) $57 - y$
10) $\frac{9}{x}$
11) $\frac{x^2}{25}$
12) $x - 6 = 19$
13) $10a - b^2$
14) $41 - ab$

Simplifying Variable Expressions

1) $3x + 15$
2) $-28x + 20$
3) $5x + 5$
4) $-8x^2 - 4$
5) $13x^2 + 10$
6) $18x^2 + 7x$
7) $-9x^2 + 4x$
8) $4x^2 - 10x$
9) $-22x + 21$
10) $68x - 12$
11) $-18x - 71$
12) $-11x^2 + 5x$
13) $-6x + 12$
14) $x - 4$
15) $22x - 5$
16) $14x - 22$
17) $27x - 16$
18) $25x + 28$
19) $-23x - 20$
20) $-3x^2 - 14x$
21) $-20x^2 + 27x$
22) $-20x^2 - 15x$
23) $4x^2 + 25x - 19$
24) $6x^2 - 66x + 25$
25) $3x^2 + 19x - 5$
26) $-7x^2 - 20x$
27) $-12x^2 + 10x - 7$
28) $-6x^2 - 21x + 13$
29) $10x^2 + 7x + 17$
30) $25x^2 + 25x$
31) $-16x^2 - 23x + 29$
32) $-9x^2 + 3x + 30$

The Distributive Property

1) $8x + 4$
2) $14x + 8$
3) $12x - 12$
4) $-12x + 30$
5) $-3x - 18$
6) $6x + 8$
7) $15x - 40$
8) $7x + 5$

9) $18x-9$
10) $-4x+28$
11) $3x-5$
12) $12x+27$
13) $18x+24$
14) $-10x+6$
15) $24x-15$
16) $-36x-36$
17) $-18x+30$
18) $24x+8$
19) $56x-24$
20) $-8x+12$
21) $45x-63$
22) $-10x+80$
23) $-12x+33$
24) $-60x+24$
25) $63x-21$
26) $-9x-81$
27) $-35x+21$
28) $50x-40$
29) $48x-96$
30) $30x-39$
31) $-20x+36$
32) $-13x-38$

Evaluating One Variables

1) 3
2) -4
3) 19
4) -17
5) 8
6) 8
7) 20
8) - 15
9) 13
10) 3
11) 14
12) 8
13) 23
14) -11
15) 35
16) 38
17) 144
18) -60
19) 32
20) 23
21) 3
22) -36
23) -66
24) 36
25) 75
26) 18
27) -74
28) -39
29) 46
30) 1
31) 6
32) 13

Combining like Terms

1) $14x+6$
2) $16x-48$
3) $11x+11$
4) $-24x+28$
5) $12x-5$
6) $40x-13$
7) $-8x+26$
8) $-35x+17$
9) $6x+1$
10) $33x-31$
11) $31x-11$
12) $-15x+20$
13) $13x-6$
14) $-18x-4$
15) $23x+44$
16) $-10x-14$
17) $37x+50$
18) $19x+17$
19) $13x+19$
20) $22x-33$
21) $x+12$
22) $-25x+30$
23) $-42x+11$
24) $-42x-27$
25) $-29x+10$
26) $-14x+14$
27) $-56x+18$
28) $27x-42$
29) $9x+15$
30) $36x-31$
31) $-15x-1$
32) $51x$

Chapter 9 :
Equations and Inequalities

Topics that you will practice in this chapter:

- ✓ One–Step Equations
- ✓ Two–Step Equations
- ✓ Multi–Step Equations
- ✓ Graphing Single–Variable Inequalities
- ✓ One–Step Inequalities
- ✓ Two–Step Inequalities
- ✓ Multi-Step Inequalities

"Life is a math equation. In order to gain the most, you have to know how to convert negatives into positives." – Anonymous

One–Step Equations

Find the answer for each equation.

1) $3x = 90, x =$ ____

2) $5x = 35, x =$ ____

3) $6x = 24, x =$ ____

4) $24x = 144, x =$ ____

5) $x + 15 = 20, x =$ ____

6) $x - 7 = 4, x =$ ____

7) $x - 9 = 2, x =$ ____

8) $x + 15 = 23, x =$ ____

9) $x - 4 = 13, x =$ ____

10) $12 = 16 + x, x =$ ____

11) $x - 10 = 2, x =$ ____

12) $5 - x = -11, x =$ ____

13) $28 = -6 + x, x =$ ____

14) $x - 20 = -35, x =$ ____

15) $x + 14 = -4, x =$ ____

16) $14 = 28 - x, x =$ ____

17) $7 + x = -7, x =$ ____

18) $x - 16 = 4, x =$ ____

19) $30 = x - 15, x =$ ____

20) $x - 5 = -18, x =$ ____

21) $x - 10 = 24, x =$ ____

22) $x - 20 = -25, x =$ ____

23) $x - 17 = 30, x =$ ____

24) $-70 = x - 28, x =$ ____

25) $x - 9 = 13, x =$ ____

26) $36 = 4x, x =$ ____

27) $x - 35 = 25, x =$ ____

28) $x - 25 = 10, x =$ ____

29) $70 - x = 16, x =$ ____

30) $x - 10 = 14, x =$ ____

31) $17 - x = -13, x =$ __

32) $x - 9 = -30, x =$ ____

One–Step Equation Word Problems

Solve.

1) How many boxes of envelopes can you buy with $40 if one box costs $5?

2) After paying $8.15 for a salad, Riya has $61.53. How much money did she have before buying the salad?

3) How many packages of Tissues can you buy with $81 if one package costs $4.5?

4) Last week Joe ran 40 miles more than Harrison. Joe ran 78 miles. How many miles did Harrison run?

5) Last Friday Liam had $65.46. Over the weekend he received some money for cleaning the attic. He now has $60. How much money did he receive?

6) After paying $17.36 for a sandwich, Elise has $31.23. How much money did she have before buying the sandwich?

Two-Steps Equations

Solve each equation.

1) $6(3 + x) = 42$

2) $(-7)(x - 2) = 56$

3) $(-8)(3x - 4) = (-16)$

4) $5(2 + x) = -15$

5) $19(3x + 11) = 38$

6) $4(2x + 2) = 24$

7) $5(8 + 3x) = (-20)$

8) $(-5)(5x - 3) = 40$

9) $2x + 12 = 16$

10) $\frac{4x - 5}{5} = 3$

11) $(-3) = \frac{x + 4}{7}$

12) $80 = (-8)(x - 3)$

13) $\frac{x}{3} + 7 = 19$

14) $\frac{1}{4} = \frac{1}{2} + \frac{x}{4}$

15) $\frac{11 + x}{5} = (-6)$

16) $(-3)(10 + 5x) = (-15)$

17) $(-3x) + 12 = 24$

18) $\frac{x+5}{5} = -5$

19) $\frac{x + 23}{8} = 3$

20) $(-4) + \frac{x}{2} = (-14)$

21) $-5 = \frac{x + 7}{8}$

22) $\frac{9x - 3}{6} = 4$

23) $\frac{2x - 12}{8} = 6$

24) $40 = (-5)(x - 8)$

Multi–Step Equations

Find the answer for each equation.

1) $3x + 3 = 9$

2) $-x + 5 = 12$

3) $4x - 8 = 8$

4) $-(3 - x) = 5$

5) $4x - 8 = 16$

6) $12x - 15 = 9$

7) $2x - 18 = 2$

8) $4x + 8 = 16$

9) $24x + 27 = 75$

10) $-14(3 + x) = 14$

11) $-3(2 + x) = 6$

12) $12 = -(x - 7)$

13) $3(3 - x) = 30$

14) $-15 = -(3x + 6)$

15) $40(3 + x) = 40$

16) $5(x - 10) = 25$

17) $-18 = x + 8x$

18) $3x + 25 = -2x - 10$

19) $7(6 + 3x) = -63$

20) $18 - 3x = -4 - 5x$

21) $4 - 6x = 36 + 2x$

22) $15 + 15x = -5 + 5x$

23) $42 = (-6x) - 7 + 7$

24) $21 = 3x - 21 + 4x$

25) $-18 = -6x - 9 + 3x$

26) $5x - 15 = -29 + 6x$

27) $7x - 18 = 4x + 3$

28) $-7 - 4x = 5(4 - x)$

29) $x - 5 = -5(-3 - x)$

30) $13x - 68 = 15x - 102$

31) $-5x - 3 = -3(9 + 3x)$

32) $-2x - 15 = 6x + 17$

One-Step Inequalities

Solve each inequality.

1) $7x < 14$

2) $x + 7 \geq -8$

3) $x - 1 \leq 9$

4) $-2x + 4 > -10$

5) $x + 18 \geq -6$

6) $x + 9 \geq 5$

7) $x - \frac{1}{3} \leq 5$

8) $-7x < 42$

9) $-x + 8 > -3$

10) $\frac{x}{3} + 3 > -9$

11) $-x + 8 > -4$

12) $x - 14 \leq 18$

13) $-x - 5 \leq -7$

14) $x + 26 \geq -13$

15) $x + \frac{1}{3} \geq -\frac{2}{3}$

16) $x + 6 \geq -14$

17) $x - 42 \leq -48$

18) $x - 5 \leq 4$

19) $-x + 5 > -6$

20) $x + 6 \geq -12$

21) $8x + 6 \leq 22$

22) $4x - 3 \geq 9$

23) $3x - 5 < 22$

24) $6x - 8 \leq 40$

Graphing Inequalities

Draw a graph for each inequality.

1) $x > -1$

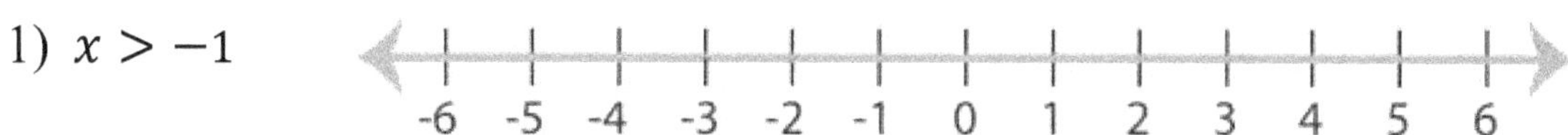

2) $x \leq 2$

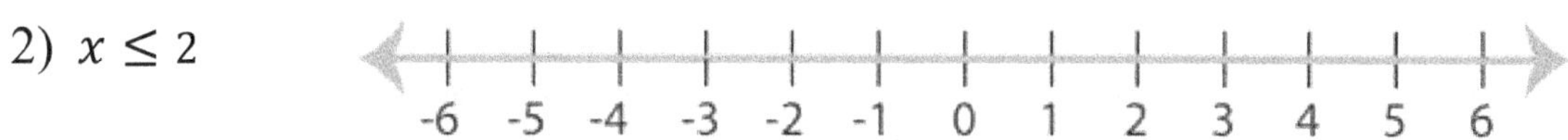

3) $x \geq 0$

-6 -5 -4 -3 -2 -1 0 1 2 3 4 5 6

4) $x < -3$

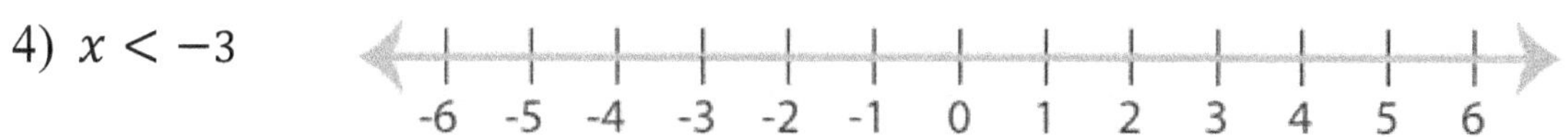

5) $x < \frac{1}{2}$

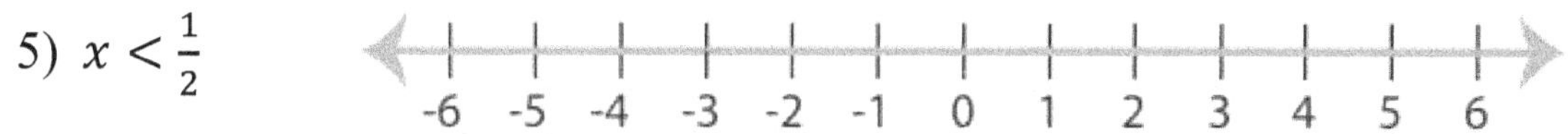

6) $x \leq -2$

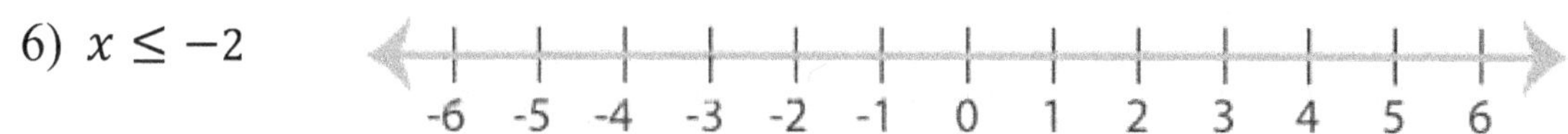

7) $x \leq 3$

-6 -5 -4 -3 -2 -1 0 1 2 3 4 5 6

8) $x \geq -\frac{7}{2}$

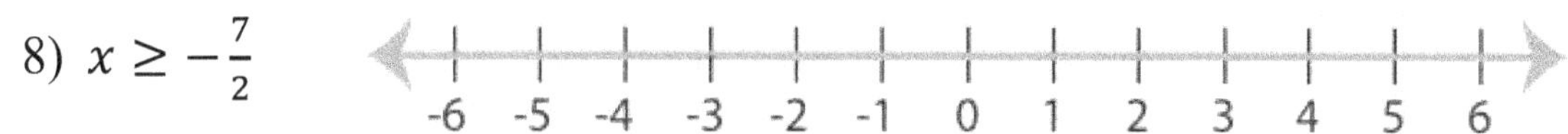

Two-Steps Inequality

Solve each inequality

1) $2x - 3 \leq 7$

2) $3x - 4 \leq 8$

3) $\frac{-1}{4}x + \frac{x}{2} \leq \frac{1}{8}$

4) $5x + 10 \geq 30$

5) $4x - 7 \geq 9$

6) $3x - 5 \leq 16$

7) $8x - 2 \leq 14$

8) $9x + 5 \leq 23$

9) $2x + 10 > 32$

10) $\frac{x}{8} + 2 \leq 4$

11) $3x + 4 \geq 37$

12) $3x - 8 < 10$

13) $6 \geq \frac{x+7}{2}$

14) $3x + 9 < 48$

15) $\frac{4+x}{5} \geq 3$

16) $16 + 4x < 36$

17) $16 > 6x - 8$

18) $5 + \frac{x}{3} < 6$

19) $-4 + 4x > 24$

20) $5 + \frac{x}{7} < 3$

Multi-Step Inequalities

Calculate each inequality.

21) $x - 3 \leq 7$

22) $8 - x \leq 8$

23) $3x - 9 \leq 9$

24) $4x - 4 \geq 8$

25) $x - 7 \geq 1$

26) $5x - 15 \leq 5$

27) $6x - 8 \leq 4$

28) $-11 + 6x \leq 12$

29) $4(x - 4) \leq 16$

30) $3x - 10 \leq 11$

31) $5x - 25 < 25$

32) $9x - 5 < 22$

33) $20 - 7x \geq -15$

34) $33 + 6x < 45$

35) $8 + 8x \geq 96$

36) $7 + 3x < 13$

37) $4x - 3 < 9$

38) $5(2 - 2x) \geq -30$

39) $-(7 + 6x) < 29$

40) $12 - 8x \geq -20$

41) $-4(x - 6) > 24$

42) $\frac{3x + 9}{6} \leq 10$

43) $\frac{4x - 10}{3} \leq 2$

44) $\frac{2x - 8}{3} > 2$

45) $8 + \frac{x}{6} < 9$

46) $\frac{9x}{7} - 4 < 5$

47) $\frac{15x + 45}{15} > 1$

48) $16 + \frac{x}{4} < 6$

Answers of Worksheets

One–Step Equations

1) 30	9) 17	17) −14	25) 22
2) 7	10) −4	18) 20	26) 9
3) 4	11) 12	19) 45	27) 60
4) 6	12) 16	20) −13	28) 35
5) 5	13) 34	21) 34	29) 54
6) 11	14) −15	22) −5	30) 24
7) 11	15) −18	23) 47	31) 30
8) 8	16) 14	24) −42	32) −21

One–Step Equation Word Problems

1) 8	3) 18	5) 5.46
2) $69.68	4) 38	6) 48.59

Two Steps Equations

1) $x = 4$	9) $x = 2$	17) $x = -4$
2) $x = -6$	10) $x = 5$	18) $x = -30$
3) $x = 2$	11) $x = -25$	19) $x = 1$
4) $x = -5$	12) $x = -7$	20) $x = -20$
5) $x = -3$	13) $x = 36$	21) $x = -47$
6) $x = 2$	14) $x = -1$	22) $x = 3$
7) $x = -4$	15) $x = -41$	23) $x = 30$
8) $x = -1$	16) $x = -1$	24) $x = 0$

Multi–Step Equations

1) 2	8) 2	15) −2	22) −2
2) −7	9) 2	16) 15	23) −7
3) 4	10) −4	17) −2	24) 6
4) 8	11) −4	18) −7	25) 3
5) 6	12) −5	19) −5	26) 14
6) 2	13) −7	20) −11	27) 7
7) 10	14) 3	21) −4	28) 27

29) -5 30) 17 31) -6 32) -4

One Step Inequality

1) $x < 2$
2) $x \geq -15$
3) $x \leq 10$
4) $x < 7$
5) $x \geq -24$
6) $x \geq -4$
7) $x \leq \frac{16}{3}$
8) $x > -6$
9) $x < 11$
10) $x > -36$
11) $x < 12$
12) $x \leq 32$
13) $x \geq 2$
14) $x \geq -39$
15) $x \geq -1$
16) $x \geq -20$
17) $x \leq -6$
18) $x \leq 9$
19) $x < 11$
20) $x \geq -18$
21) $x \leq 2$
22) $x \geq 3$
23) $x < 9$
24) $x \leq 8$

Graphing Single–Variable Inequalities

1)

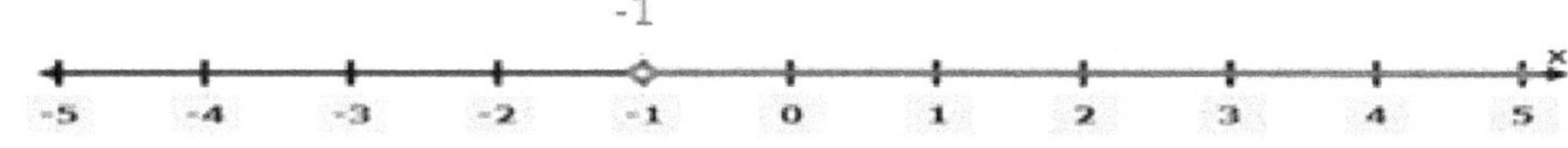

2)

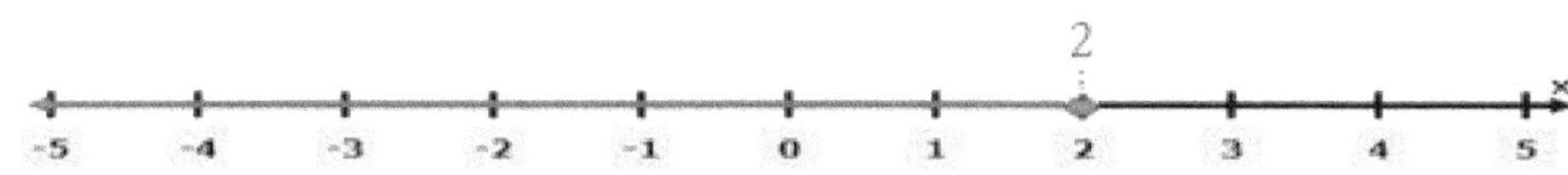

3)

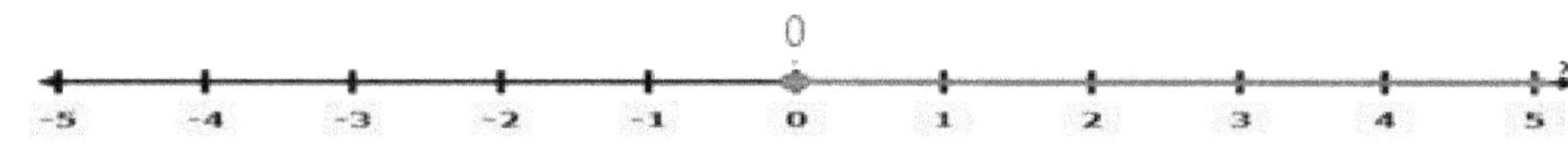

4)

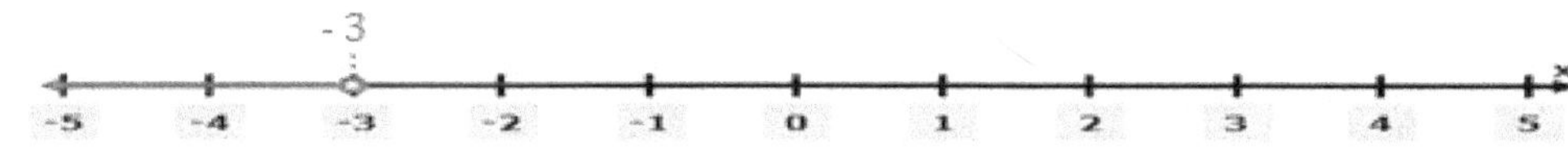

5)

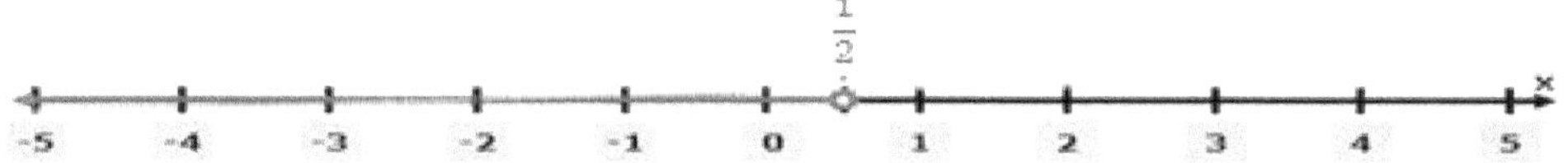

6)

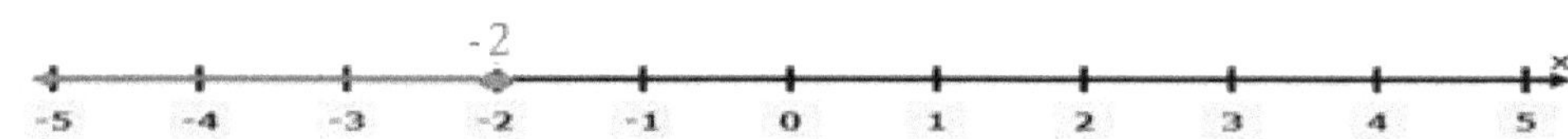

7)

8)

Two Steps Inequality

1) $x \leq 5$
2) $x \leq 4$
3) $x \leq 0.5$
4) $x \geq 4$
5) $x \geq 4$
6) $x \leq 7$

7) $x \leq 2$
8) $x \leq 2$
9) $x > 11$
10) $x \leq 16$
11) $x \geq 11$
12) $x < 6$
13) $x \leq 5$
14) $x < 13$
15) $x \geq 11$
16) $x < 5$
17) $x < 4$
18) $x < 3$
19) $x > 7$
20) $x < -14$

Multi-Step Inequalities

1) $x \leq 10$
2) $x \geq 0$
3) $x \leq 6$
4) $x \geq 3$
5) $x \geq 8$
6) $x \leq 4$
7) $x \leq 2$
8) $x \leq \frac{23}{6}$
9) $x \leq 8$
10) $x \leq 7$
11) $x < 10$
12) $x < 3$
13) $x \leq 5$
14) $x < 2$
15) $x \geq 11$
16) $x < 2$
17) $x < 3$
18) $x \leq 4$
19) $x > -6$
20) $x \leq 4$
21) $x < 0$
22) $x \leq 17$
23) $x \leq 4$
24) $x > 7$
25) $x < 6$
26) $x < 7$
27) $x > -2$
28) $x < -40$

Chapter 10 :
Geometry and Solid Figures

Topics that you will practice in this chapter:

- ✓ Angles
- ✓ Pythagorean Relationship
- ✓ Triangles
- ✓ Polygons
- ✓ Trapezoids
- ✓ Circles
- ✓ Cubes
- ✓ Rectangular Prism
- ✓ Cylinder

Mathematics is, as it were, a sensuous logic, and relates to philosophy as do the arts, music, and plastic art to poetry. — K. Shegel

Angles

What is the value of x in the following figures?

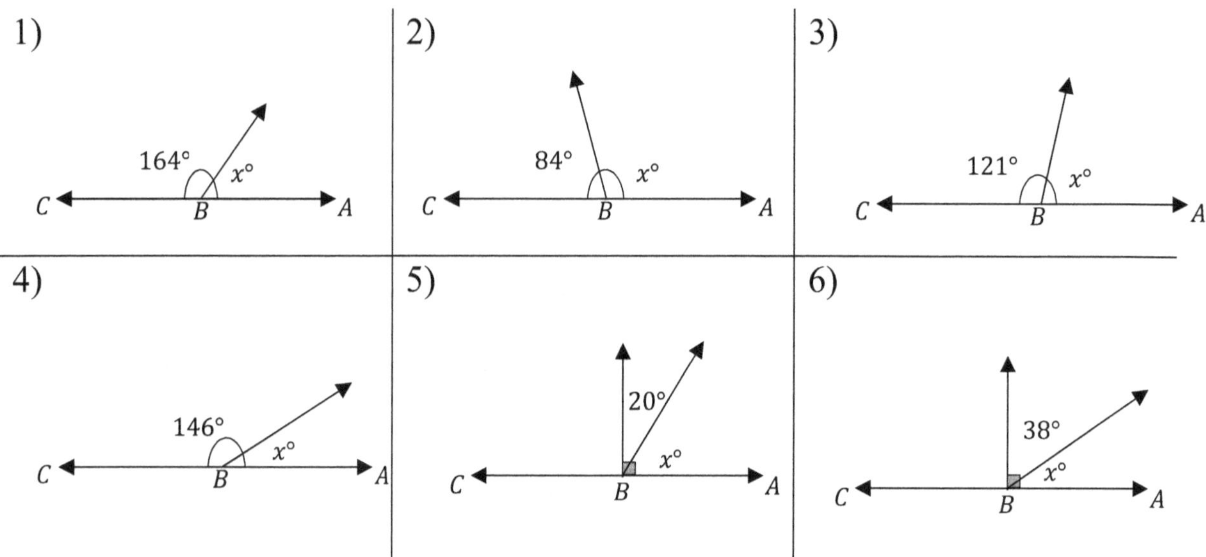

Calculate.

7) Two supplement angles have equal measures. What is the measure of each angle? ____________________

8) The measure of an angle is seven fifth the measure of its supplement. What is the measure of the angle? ____________________

9) Two angles are complementary and the measure of one angle is 24 less than the other. What is the measure of the smaller angle? ____________________

10) Two angles are complementary. The measure of one angle is one fifth the measure of the other. What is the measure of the bigger angle? ____________________

11) Two supplementary angles are given. The measure of one angle is 40° less than the measure of the other. What does the smaller angle measure? ____________________

Pythagorean Relationship

Do the following lengths form a right triangle?

1)

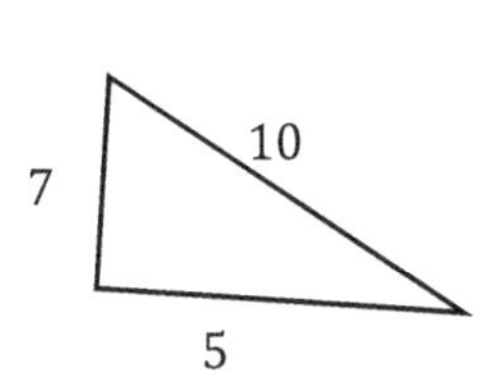

2)

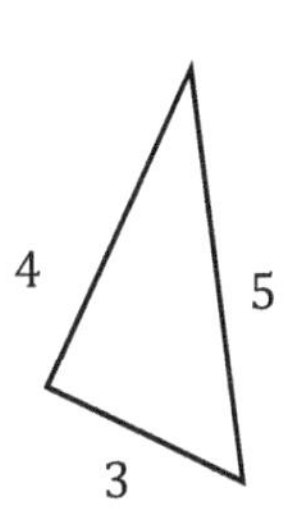

3)

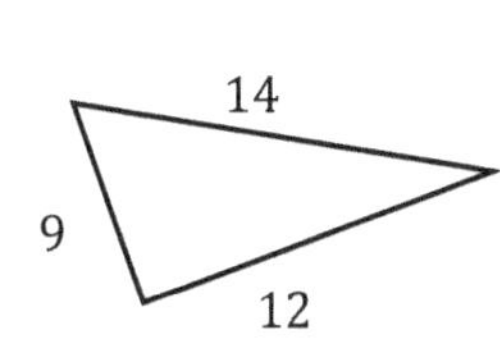

4)

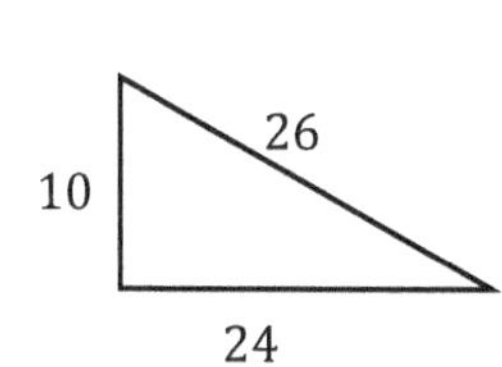

5)

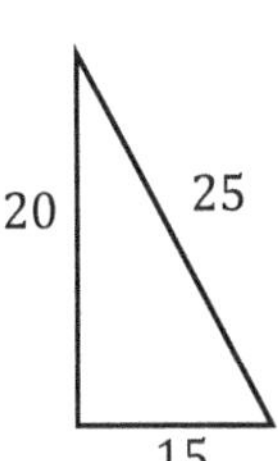

6)

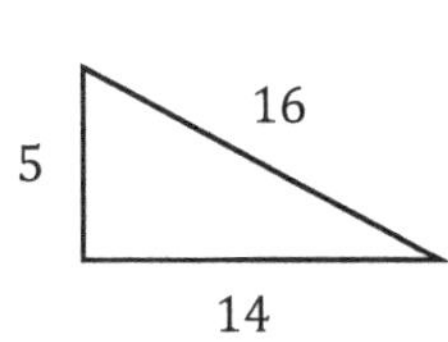

7)

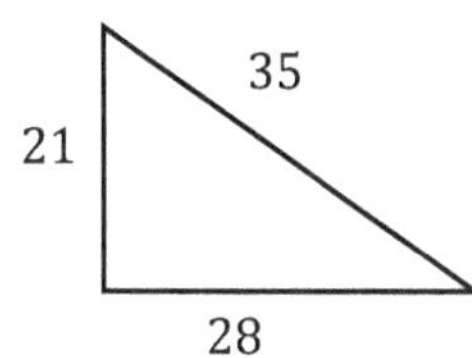

8)

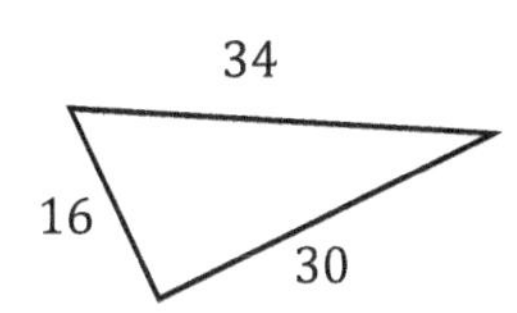

Find the missing side?

9)

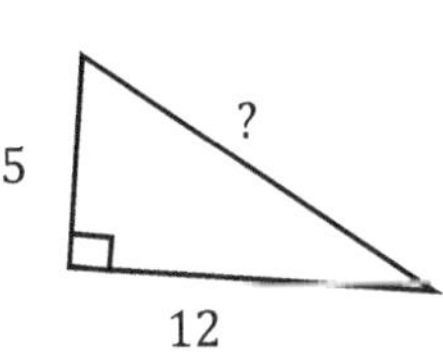

10)

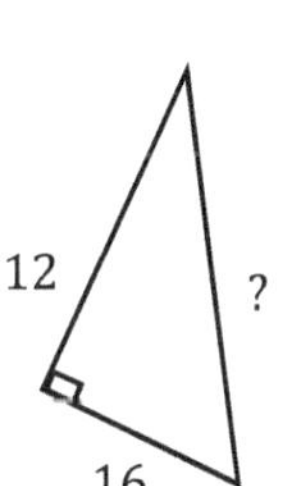

11)

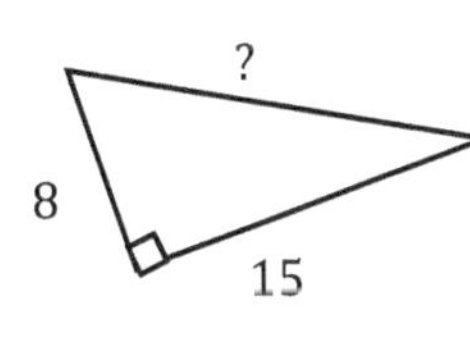

12)

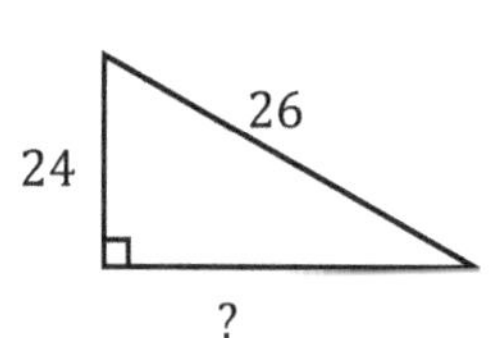

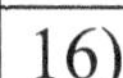

13)

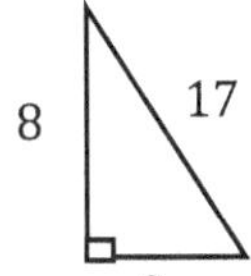

14)

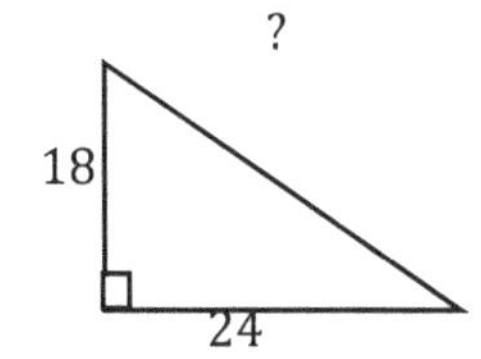

15)

15
39
?

16)

15
9
?

Triangles

Find the measure of the unknown angle in each triangle.

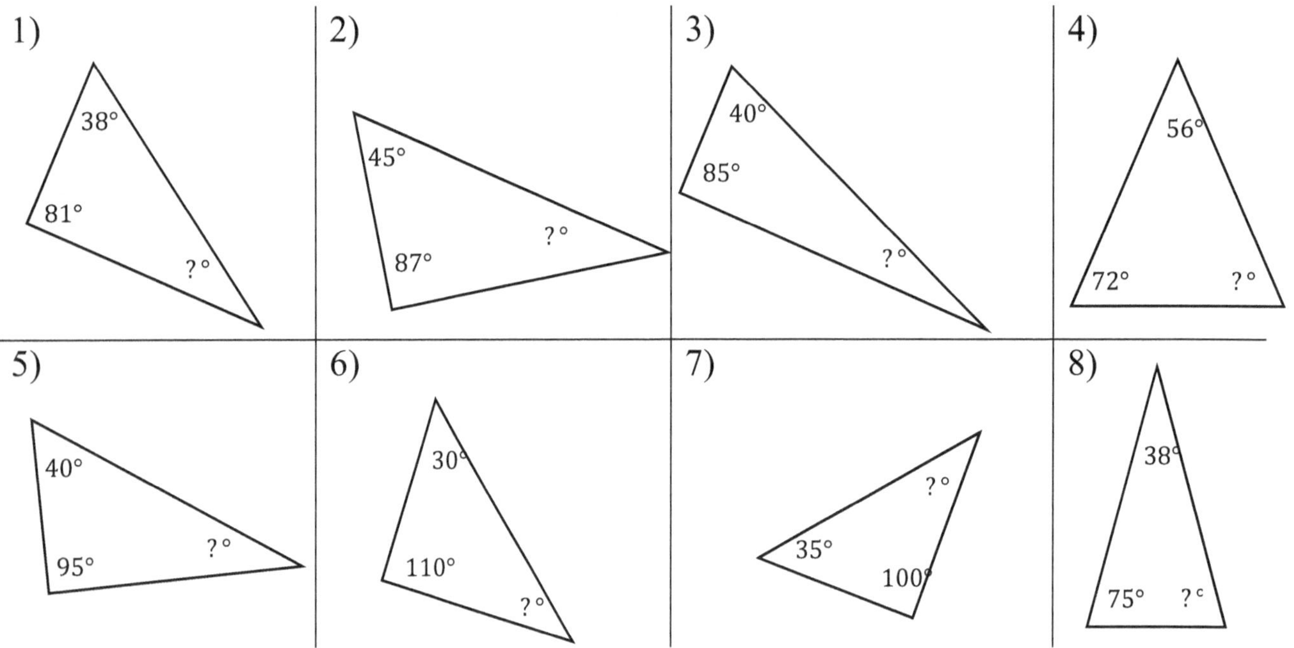

Find area of each triangle.

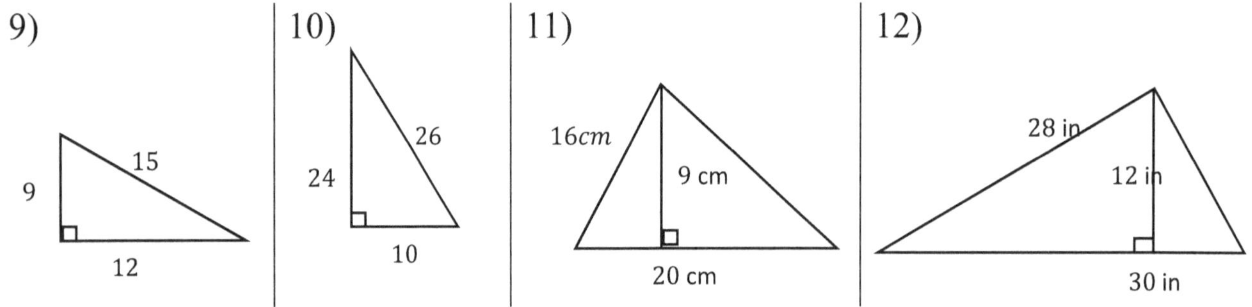

Polygons

Find the perimeter of each shape.

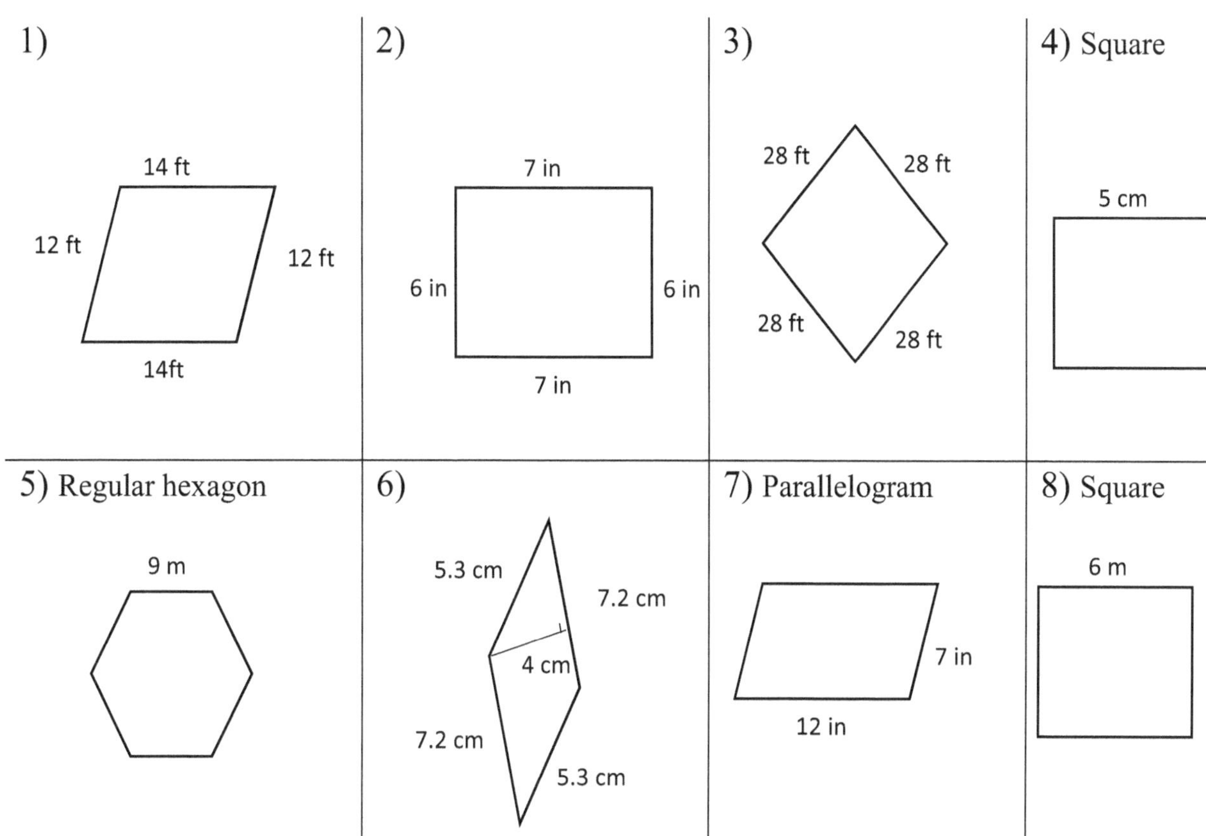

Find the area of each shape.

9) Parallelogram

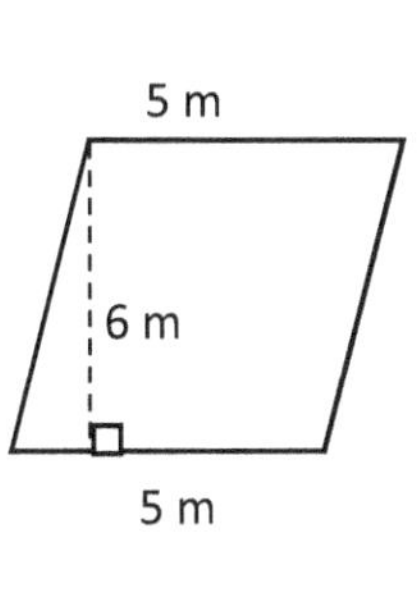

10) Rectangle

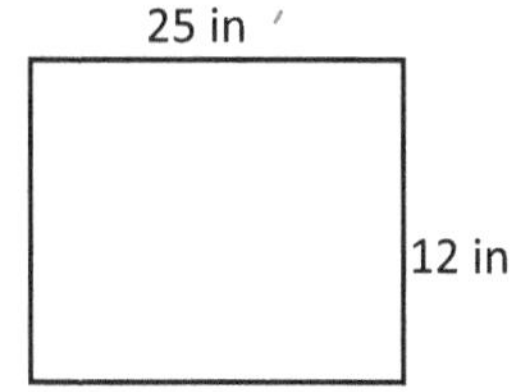

11) Rectangle

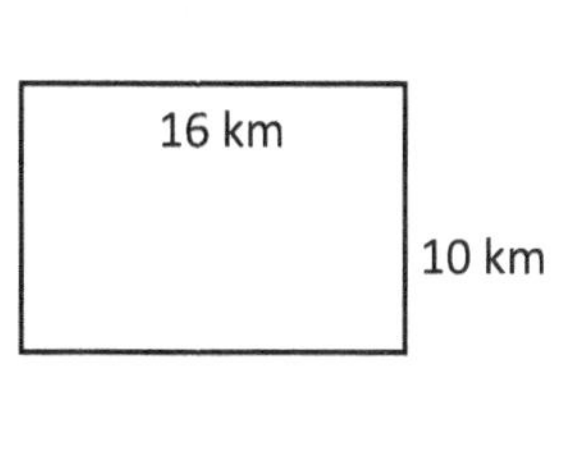

12) Square

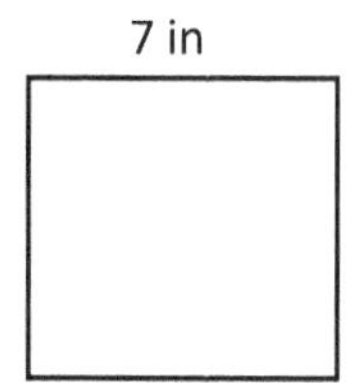

Trapezoids

Find the area of each trapezoid.

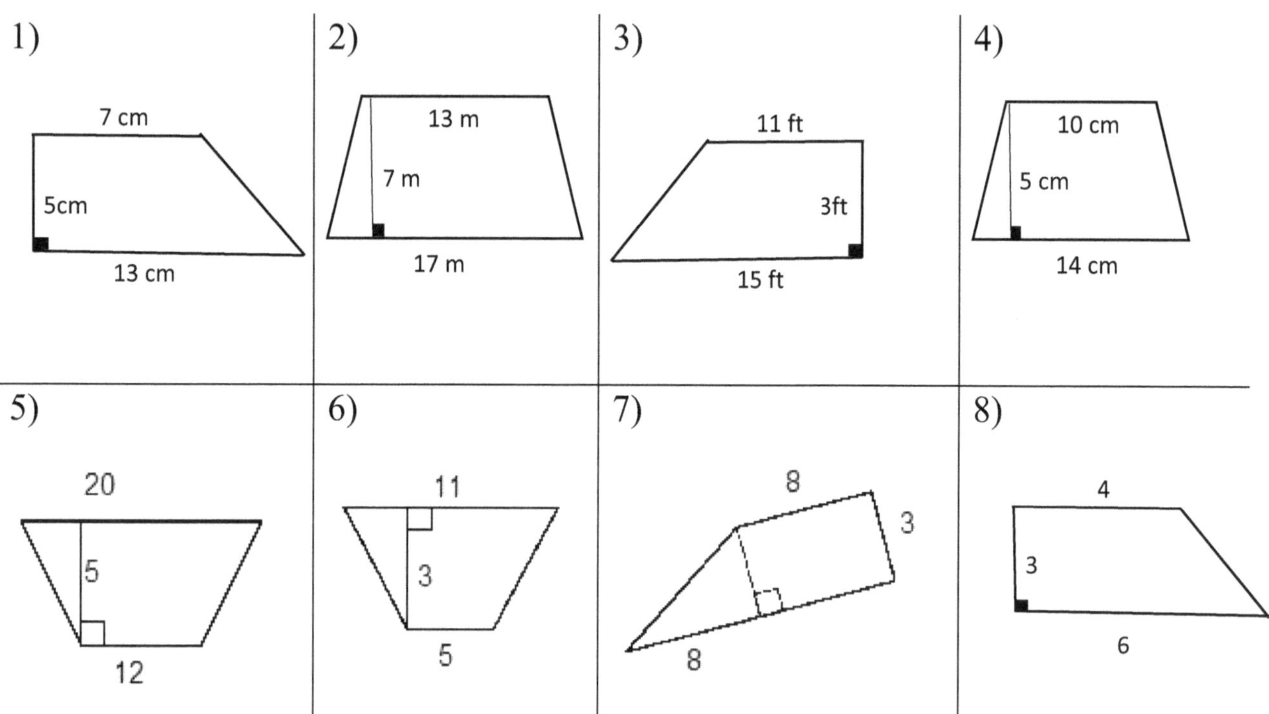

Calculate.

1) A trapezoid has an area of 45 cm^2 and its height is 5 cm and one base is 5 cm. What is the other base length? ____________________

2) If a trapezoid has an area of 99 ft^2 and the lengths of the bases are 8 ft and 10 ft, find the height? ____________________

3) If a trapezoid has an area of 126 m^2 and its height is 14 m and one base is 6 m, find the other base length? ____________________

4) The area of a trapezoid is 440 ft^2 and its height is 22 ft. If one base of the trapezoid is 15 ft, what is the other base length?

____

Circles

Find the area of each circle. ($\pi = 3.14$)

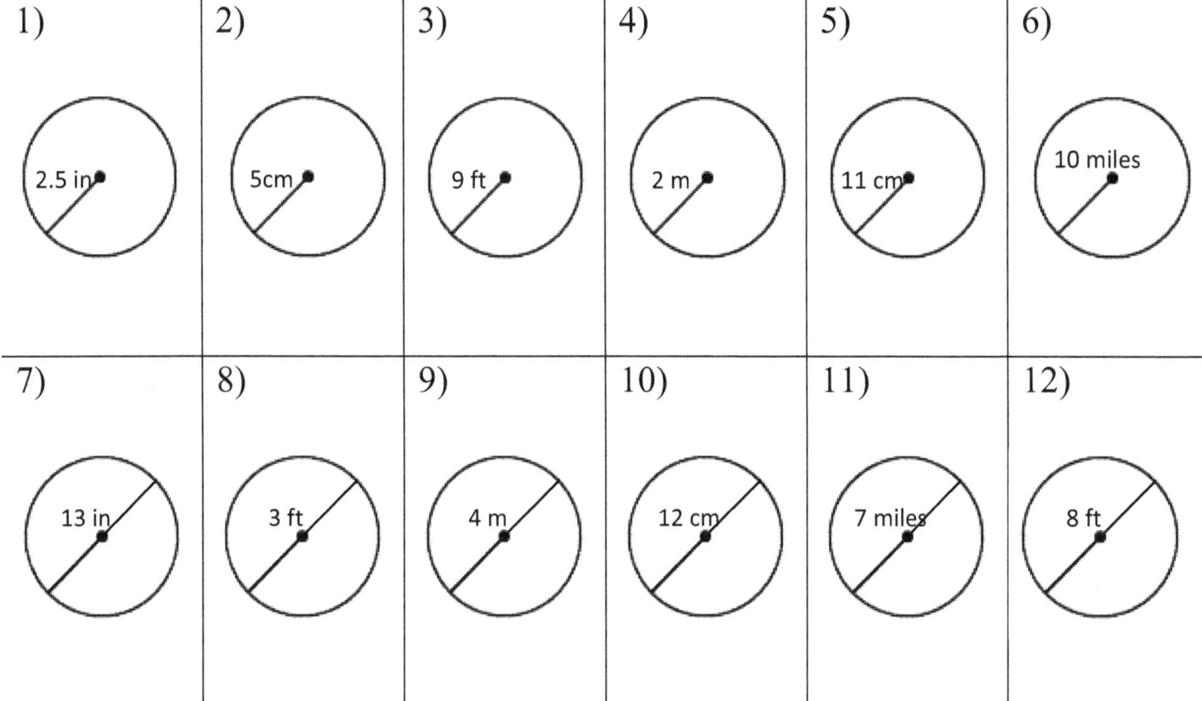

Complete the table below. ($\pi = 3.14$)

Circle No.	Radius	Diameter	Circumference	Area
1	$1\ in$	$2\ in$	$6.28\ in$	$3.14\ in^2$
2		$10\ m$		
3				$28.26\ ft^2$
4			$47.1\ mi$	
5		$11\ km$		
6	$7\ cm$			
7		$12\ ft$		
8				$314\ m^2$
9			$56.52\ in$	
10	$4.5\ ft$			

Cubes

Find the volume of each cube.

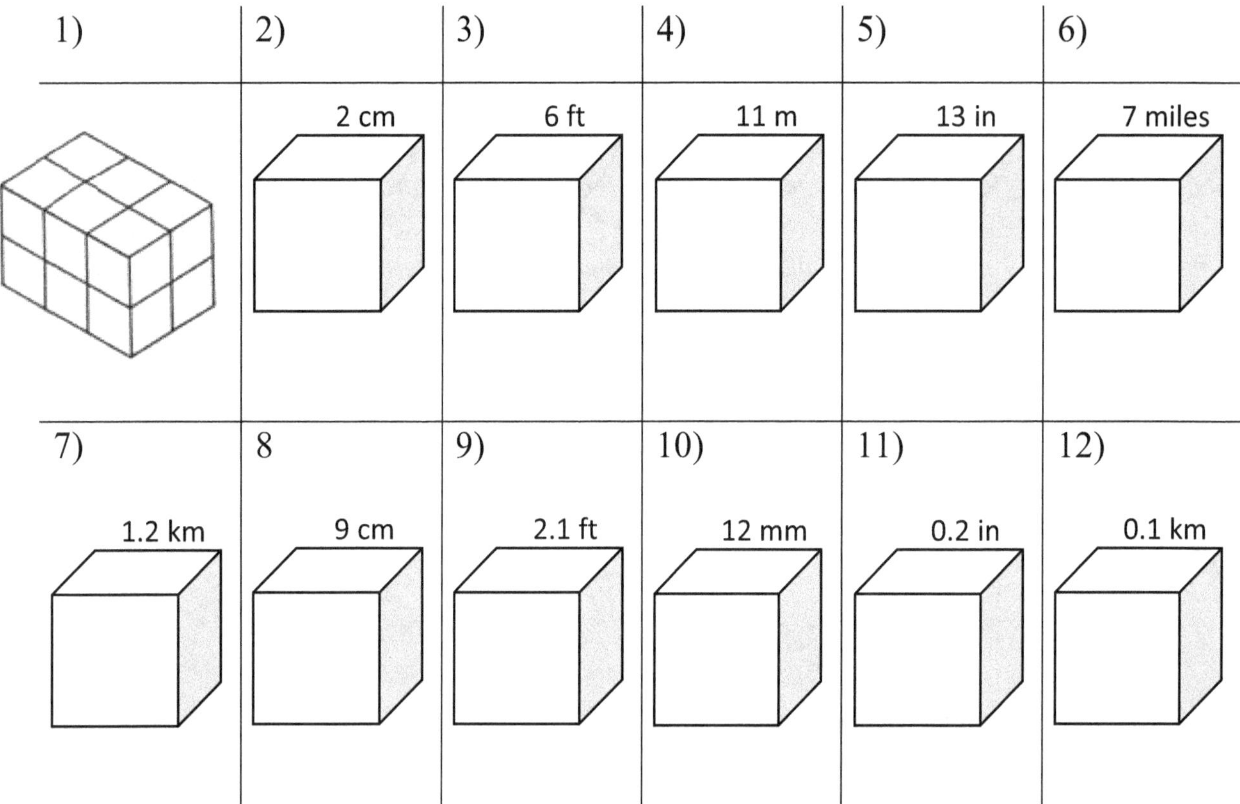

Find the surface area of each cube.

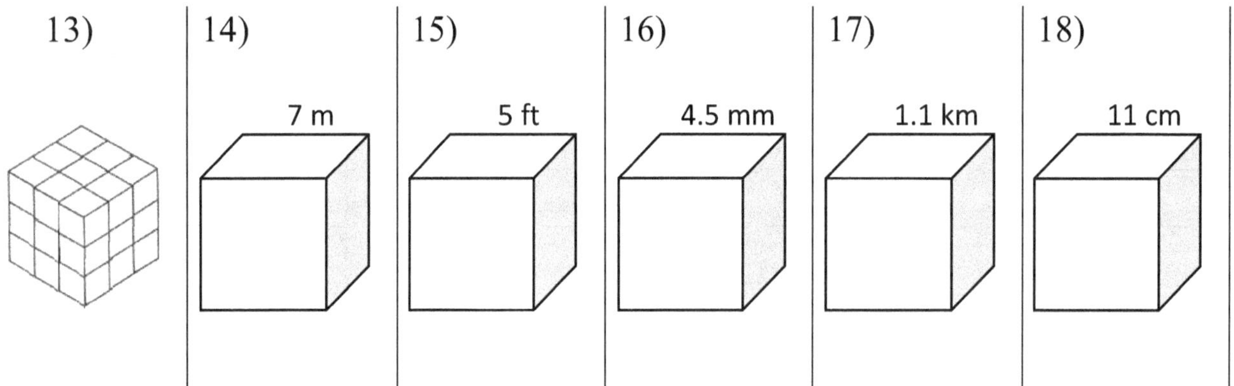

Rectangular Prism

Find the volume of each Rectangular Prism.

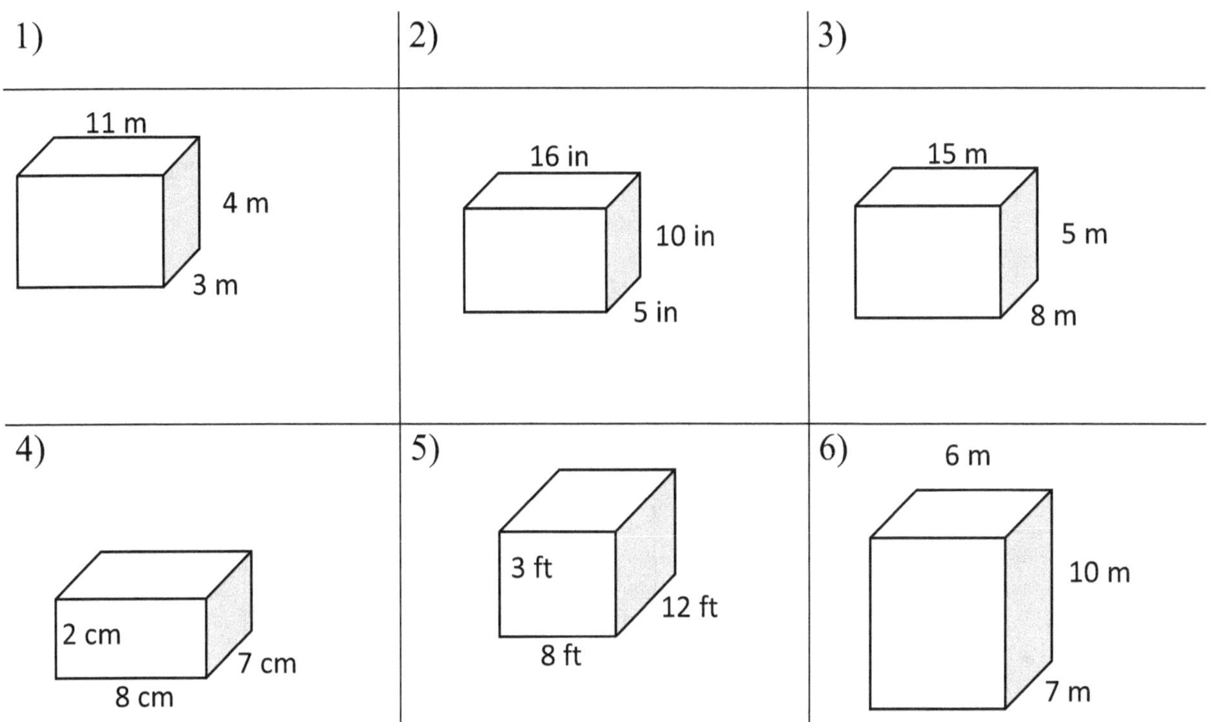

Find the surface area of each Rectangular Prism.

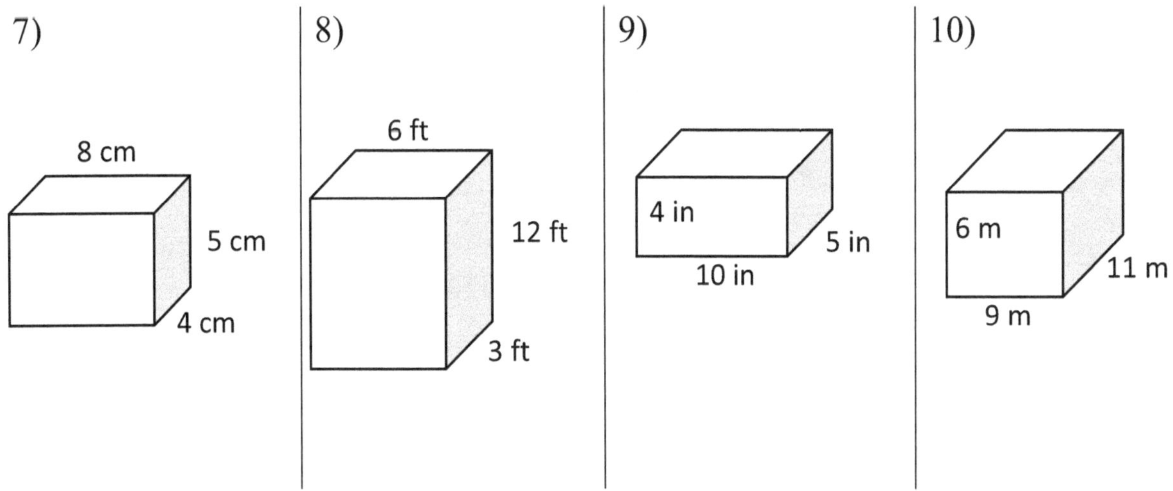

Cylinder

Find the volume of each Cylinder. Round your answer to the nearest tenth. ($\pi = 3.14$)

1)

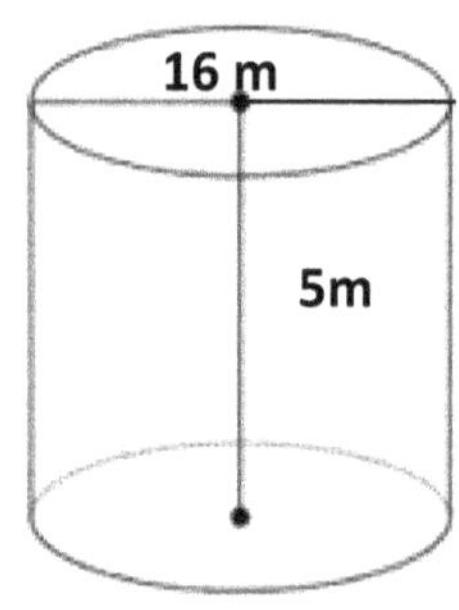

2)

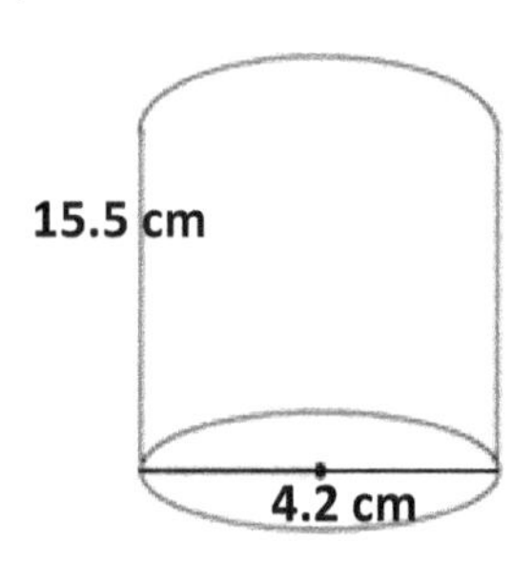

3)

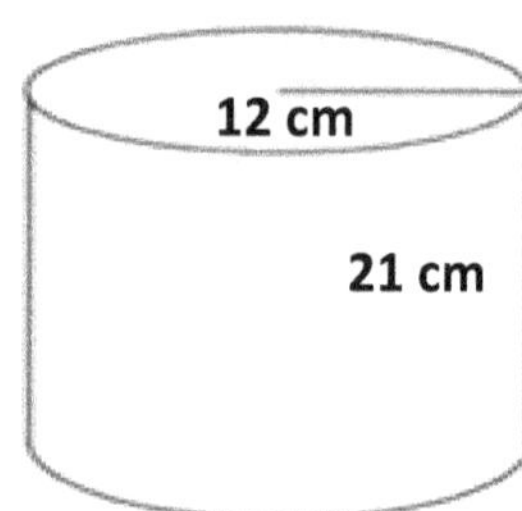

4)

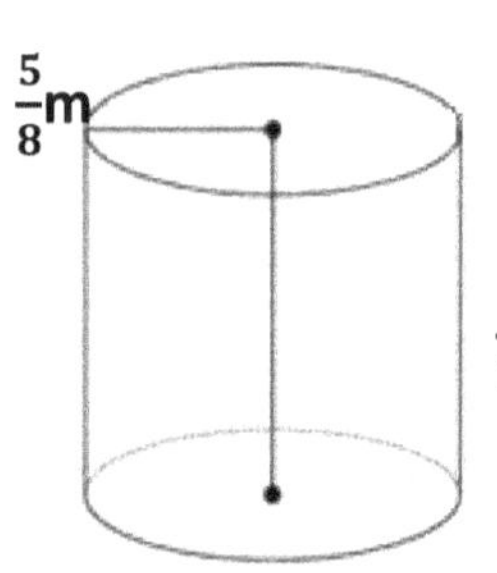

5)

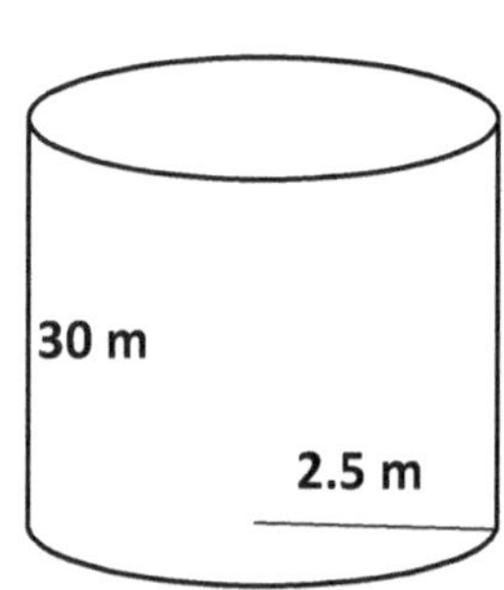

6)

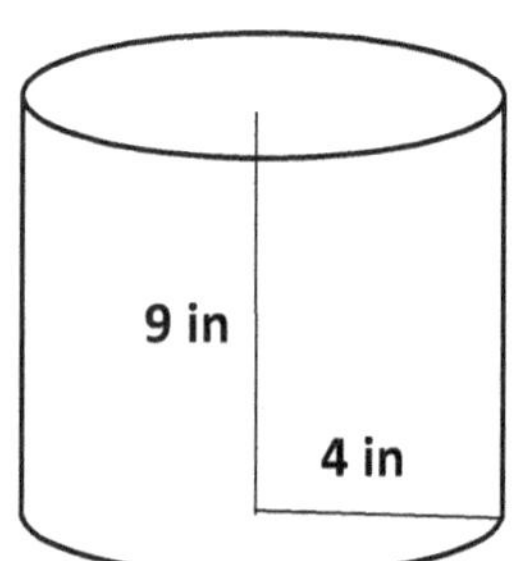

Find the surface area of each Cylinder. ($\pi = 3.14$)

7)

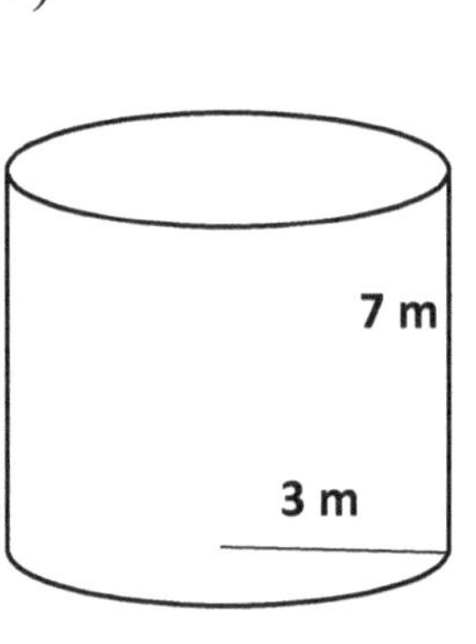

8)

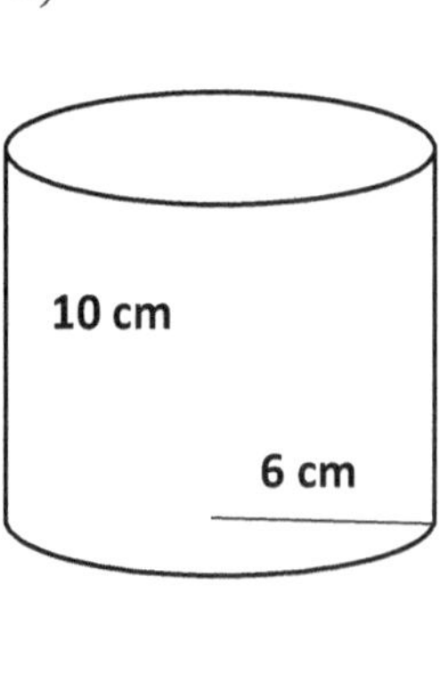

9)

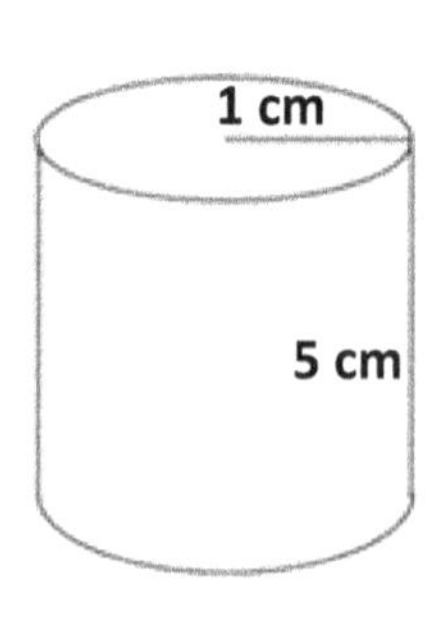

10)

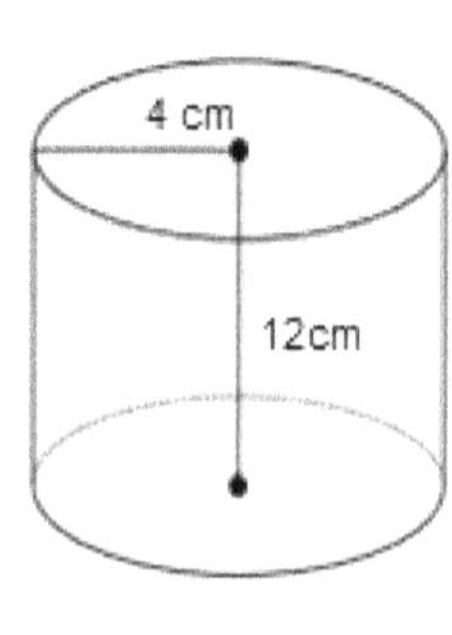

Answers of Worksheets

Angles

1) 16°
2) 96°
3) 59°
4) 34°
5) 70°
6) 52°
7) 90°
8) 75°
9) 33°
10) 75°
11) 70°

Pythagorean Relationship

1) No
2) Yes
3) No
4) Yes
5) Yes
6) No
7) Yes
8) Yes
9) 13
10) 20
11) 17
12) 10
13) 15
14) 30
15) 36
16) 12

Triangles

1) 60°
2) 48°
3) 55°
4) 52°
5) 45°
6) 40°
7) 45°
8) 67°
9) 54 *square unites*
10) 120 *square unites*
11) 90 *square unites*
12) 180 *square unites*

Polygons

1) 52 ft
2) 26 in
3) 112 ft
4) 20 cm
5) 54 m
6) 25 cm
7) 38 in
8) 24 m
9) 30 m^2
10) 300 in^2
11) 160 km^2
12) 49 in^2

Trapezoids

1) 50 cm^2
2) 105 m^2
3) 39 ft^2
4) 60 cm^2
5) 80
6) 24
7) 36
8) 15

Calculate

1) 13 cm
2) 11 ft
3) 12 m
4) 25 ft

Circles

1) 19.63 in^2
2) 78.5 cm^2
3) 254.34 ft^2
4) 12.56 m^2
5) 379.94 cm^2
6) 314 $miles^2$
7) 132.67 in^2
8) 7.07 ft^2
9) 12.56 m^2
10) 113.04 cm^2
11) 38.47 $miles^2$
12) 50.24 ft^2

Circle No.	Radius	Diameter	Circumference	Area
1	1 in	2 in	6.28 in	3.14 in^2
2	5 m	10 m	31.4 m	78.5 m^2
3	3 ft	6 ft	18.84 ft	28.26 ft^2
4	7.5 miles	15 mi	47.1 mi	176.63 mi^2
5	5.5 km	11 km	34.54 km	94.99 km^2
6	7 cm	14 cm	43.96 cm	153.86 cm^2
7	6 ft	12 ft	37.68 feet	113.04 ft^2
8	10 m	20 m	62.8 m	314 m^2
9	9 in	18 in	56.52 in	254.34 in^2
10	4.5 ft	9 ft	28.26 ft	63.585 ft^2

Cubes

1) 12
2) 8 cm^3
3) 216 ft^3
4) 1,331 m^3
5) 2,197 in^3
6) 343 $miles^3$
7) 1.728 km^3
8) 729 cm^3
9) 9.261 ft^3
10) 1,728 mm^3
11) 0.008 in^3
12) 0.001 km^3
13) 27
14) 294 m^2
15) 150 ft^2
16) 121.5 mm^2
17) 7.26 km^2
18) 726 cm^2

Rectangular Prism

1) 132 m^3
2) 800 in^3
3) 600 m^3
4) 112 cm^3
5) 288 ft^3
6) 420 m^3
7) 184 cm^2
8) 252 ft^2
9) 220 in^2
10) 438 m^2

Cylinder

1) 1,004.8 m^3
2) 214.6 cm^3
3) 9,495.4 cm^3
4) 1.1 m^3
5) 588.8 m^3
6) 452.2 in^3
7) 188.4 m^2
8) 602.9 cm^2
9) 37.7 cm^2
10) 401.9 m^2

Chapter 11 :

Statistics and Probability

Topics that you will practice in this chapter:

- ✓ Mean and Median
- ✓ Mode and Range
- ✓ Times Series
- ✓ Stem–and–Leaf Plot
- ✓ Pie Graph
- ✓ Probability Problems

Mathematics is no more computation than typing is literature.
– John Allen Paulos

Mean and Median

Find Mean and Median of the Given Data.

1) 8, 7, 14, 4, 8

2) 14, 8, 25, 19, 16, 33, 11

3) 23, 18, 15, 12, 17

4) 34, 14, 10, 15, 6, 11

5) 10, 19, 6, 8, 32, 20, 17

6) 17, 26, 39, 69, 20, 6

7) 40, 38, 18, 11, 9, 2, 7, 32,41

8) 24, 21, 31,12,33, 32, 22

9) 16, 14, 20, 41, 15, 20, 38, 4

10) 20, 20, 30, 18, 6, 28, 12, 46

11) 12, 7, 10, 11, 16, 22

12) 10, 29, 27, 12, 2, 15, 10, 3

Calculate.

13) In a javelin throw competition, five athletics score 56, 34, 62, 23 and 19 meters. What are their Mean and Median? ________________

14) Eva went to shop and bought 8 apples, 14 peaches, 6 bananas, 4 pineapples and 12 melons. What are the Mean and Median of her purchase? ________________

15) Bob has 17 black pen, 19 red pen, 14 green pens, 20 blue pens and 5 boxes of yellow pens. If the Mean and Median are 19 respectively, what is the number of yellow pens in each box? ________________

Mode and Range

Find Mode and Rage of the Given Data.

1) 4, 3, 7, 3, 3, 4

Mode: _____ Range: _____

2) 18, 18, 24, 26, 18, 8, 14, 22

Mode: _____ Range: _____

3) 8, 8, 8, 16, 19, 22, 20, 9, 13

Mode: _____ Range: _____

4) 24, 24, 14, 28, 20, 18, 20, 24

Mode: _____ Range: _____

5) 6, 21, 27, 24, 27, 27

Mode: _____ Range: _____

6) 21, 8, 8, 7, 8, 12, 10, 22, 18, 13

Mode: _____ Range: _____

7) 7, 4, 4, 6, 13, 13, 13, 0, 2, 2

Mode: _____ Range: _____

8) 5, 8, 5, 14, 12, 14, 3, 5, 18

Mode: _____ Range: _____

9) 7, 7, 7, 12, 7, 3, 8, 16, 3, 17

Mode: _____ Range: _____

10) 15, 15, 19, 16, 4, 16, 10, 15

Mode: _____ Range: _____

11) 6, 6, 5, 6, 42, 13, 19, 2

Mode: _____ Range: _____

12) 8, 8, 9, 8, 9, 4, 34, 22

Mode: _____ Range: _____

Calculate.

13) A stationery sold 12 pencils,56 red pens,24 blue pens,20 notebooks, 12 erasers, 21 rulers and 11 color pencils. What are the Mode and Range for the stationery sells?

Mode: _____ Range: _____

14) In an English test, eight students score 10, 15, 15, 18 18, 16, 15 and 15. What are their Mode and Range? _______________

15) What is the range of the first 6 even numbers greater than 8?

____

Times Series

Use the following Graph to complete the table.

Day	Distance (km)
1	
2	

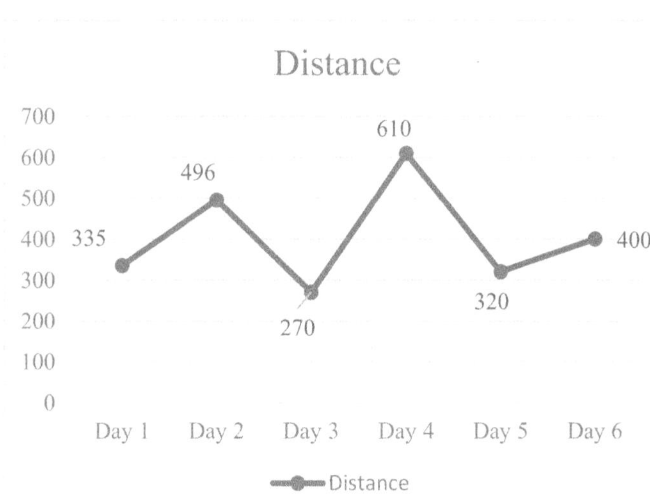

The following table shows the number of births in the US from 2007 to 2012 (in millions).

Year	Number of births (in millions)
2007	4.15
2008	3.70
2009	3.45
2010	3.20
2011	1.75
2012	2.98

Draw a Time Series for the table.

Stem–and–Leaf Plot

Make stem ad leaf plots for the given data.

1) 24, 26,29, 20, 53, 27, 51, 55, 36, 21, 37, 30

Stem | Leaf plot

2) 11, 59, 66, 14, 18, 19, 59, 65, 69, 61, 68, 65

Stem | Leaf plot

3) 121, 55, 66, 54, 112, 128, 63, 125, 59, 123, 68, 119

Stem | Leaf plot

4) 51, 32, 100, 56, 84, 36, 107, 56, 85, 39, 56, 106, 89

Stem | Leaf plot

5) 33, 89, 19, 87, 81, 16, 11, 30, 86, 35, 17, 35, 13

Stem | Leaf plot

6) 60, 92, 22, 25, 67, 93, 95, 62, 21, 64, 98, 29

Stem | Leaf plot

Quartile of a Data Set

Find First, Second and Third Quartile of the Given Data.

1) 45, 8, 25, 43, 24, 36, 35, 62, 19

2) 23, 63, 25, 19, 80, 32

3) 86, 33, 85, 60, 72, 42, 51, 46

4) 24, 44, 48, 25, 25, 36, 25, 36, 71, 49

5) 23, 15, 45, 9, 35, 8, 25, 15

6) 66, 86, 40, 32, 82, 25, 52, 44, 61

Box and Whisker Plots

Make box and whisker plots for the given data.

1) 86, 65, 92, 67, 72, 87, 87, 83, 95, 66, 76, 82

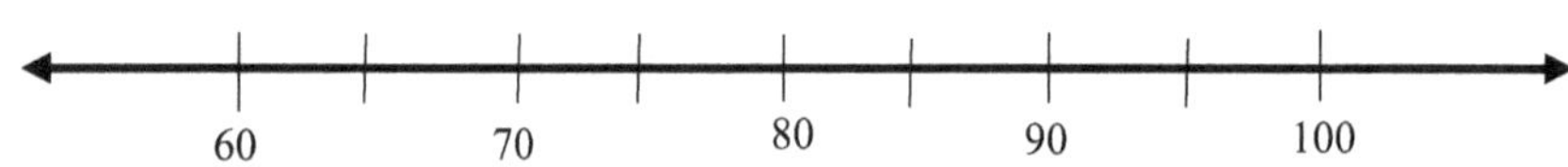

2) 8, 22, 17, 15, 13, 5, 8, 12, 6, 11, 6, 15, 4, 28

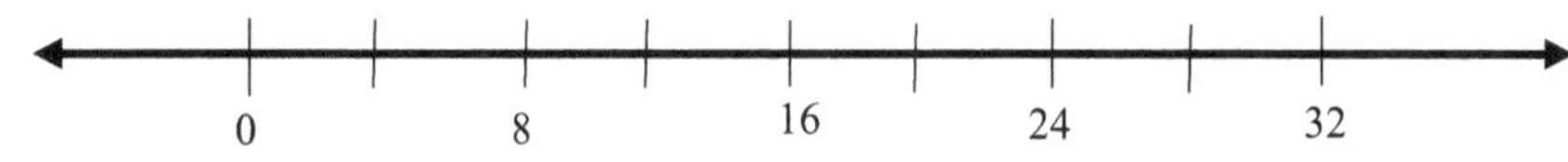

3) 25, 21, 34, 19, 23, 24,13, 17, 15, 16, 22

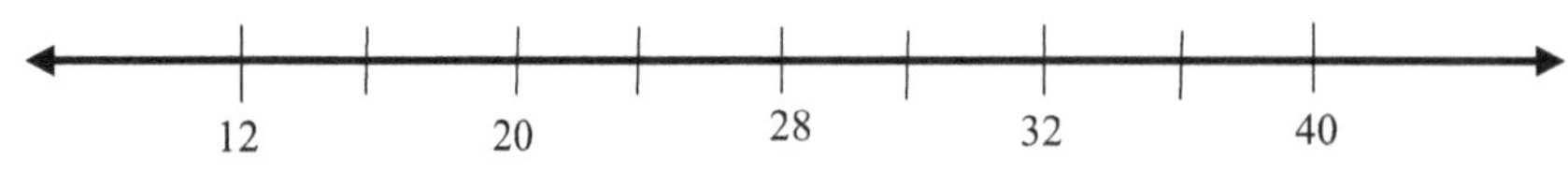

Pie Graph

The circle graph below shows all Robert's expenses for last month. Robert spent $140 on his hobbies last month.

Answer following questions based on the Pie graph.

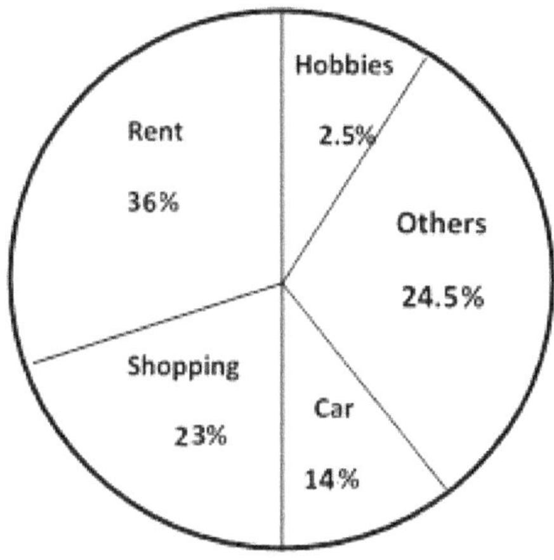

1) How much was Robert's total expenses last month? ____________________

2) How much did Robert spend on his car last month? ____________________

3) How much did Robert spend for shopping last month? ___________

4) How much did Robert spend on his rent last month? ____________________

5) What fraction is Robert's expenses for his rent and car out of his total expenses last month? ______________________

Probability Problems

Calculate.

1) A number is chosen at random from 1 to 10. Find the probability of selecting number 6 or smaller numbers. ____________________

2) Bag A contains 18 red marbles and 6 green marbles. Bag B contains 16 black marbles and 8 orange marbles. What is the probability of selecting a green marble at random from bag A? What is the probability of selecting a black marble at random from Bag B? ____________________

3) A number is chosen at random from 1 to 20. What is the probability of selecting multiples of 4? ____________________

4) A card is chosen from a well-shuffled deck of 52 cards. What is the probability that the card will be a queen? ____________________

5) A number is chosen at random from 1 to 15. What is the probability of selecting a multiple of 3 or 5? ____________________

A spinner numbered 1–8, is spun once. What is the probability of spinning ...?

6) an Odd number? __________ 7) a multiple of 2? _____

8) a multiple of 5? _____ 9) number 10? ________

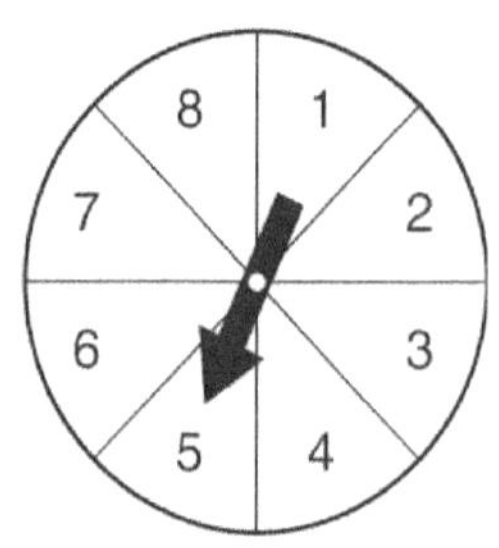

Answers of Worksheets

Mean and Median

1) Mean: 8.2, Median: 8
2) Mean: 18, Median: 16
3) Mean: 17, Median: 17
4) Mean: 15, Median: 12.5
5) Mean: 16, Median: 17
6) Mean: 29.5, Median: 23
7) Mean: 22, Median: 18
8) Mean: 25, Median: 24
9) Mean: 21, Median: 18
10) Mean: 22.5, Median: 20
11) Mean: 13, Median: 11.5
12) Mean: 13.5, Median: 11
13) Mean: 38.8, Median: 34
14) Mean: 8.8, Median: 8
15) 5

Mode and Range

1) Mode: 3, Range: 4
2) Mode: 18, Range: 18
3) Mode: 8, Range: 14
4) Mode: 24, Range: 14
5) Mode: 27, Range: 21
6) Mode: 8, Range: 15
7) Mode: 13, Range: 13
8) Mode: 5, Range: 15
9) Mode: 7, Range: 14
10) Mode: 15, Range: 15
11) Mode: 6, Range: 40
12) Mode: 8, Range: 30
13) Mode: 12, Range: 45
14) Mode: 15, Range: 8
15) 10

Time series

Day	Distance (km)
1	335
2	496
3	270
4	610
5	320
6	400

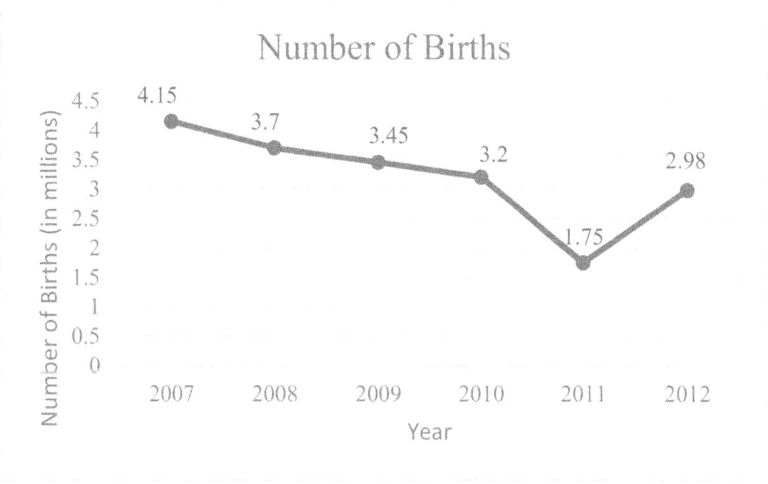

Stem–And–Leaf Plot

1)

Stem	leaf
2	0 1 4 6 7 9
3	0 6 7
5	1 3 5

2)

Stem	leaf
1	1 4 8 9
5	9 9
6	1 5 5 6 8 9

3)

Stem	leaf
5	4 5 9
6	3 6 8
11	2 9
12	1 3 5 8

4)

Stem	leaf
3	2 6 9
5	1 6 6 6
8	4 5 9
10	0 6 7

5)

Stem	leaf
1	1 3 6 7 9
3	0 3 5 5
8	1 6 7 9

6)

Stem	leaf
2	2 1 5 9
6	0 2 4 7
9	2 3 5 8

Quartile of the Given Data

1) First quartile: 21.5, second quartile: 35, third quartile: 44
2) First quartile: 22, second quartile: 28.5, third quartile: 67.25
3) First quartile: 43, second quartile: 55.5, third quartile: 81.75
4) First quartile: 25, second quartile: 36, third quartile: 48.25
5) First quartile: 10.5, second quartile: 19, third quartile: 32.5
6) First quartile: 36, second quartile: 52, third quartile: 74

Box and Whisker Plots

1)

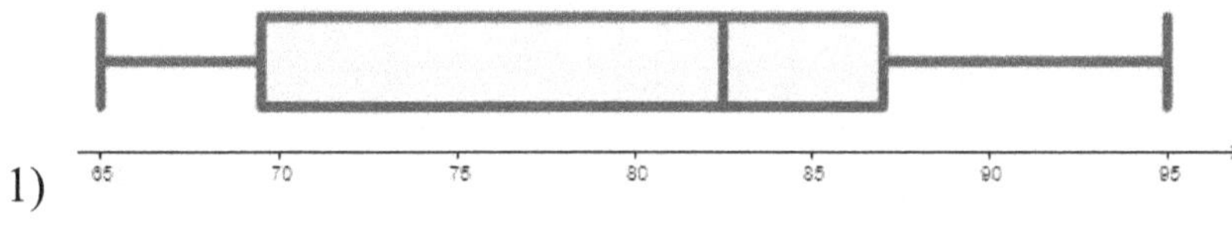

2)

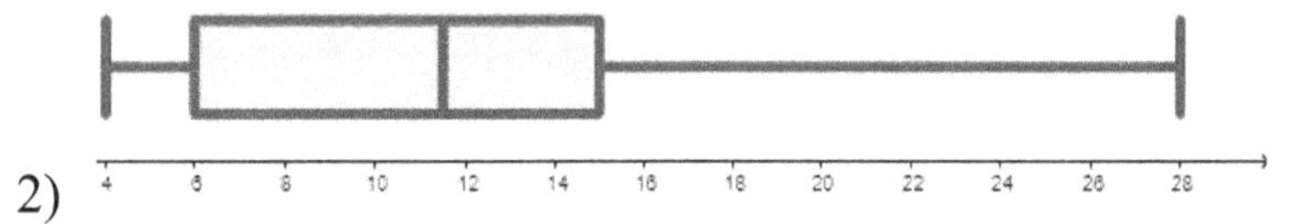

3)

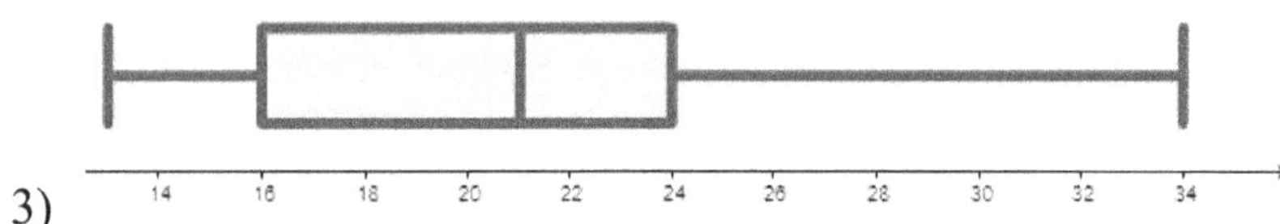

Pie Graph

1) $5,600
2) $784
3) $1,288
4) $2,016
5) $\frac{1}{2}$

Probability Problems

1) $\frac{3}{5}$
2) $\frac{1}{4}, \frac{2}{3}$
3) $\frac{1}{4}$
4) $\frac{1}{13}$
5) $\frac{7}{15}$
6) $\frac{1}{2}$
7) $\frac{1}{2}$
8) $\frac{1}{8}$
9) 0

Chapter 12 : SBAC Math Practice Tests

Smarter Balanced Assessment Consortium (SBAC) test assesses student mastery of the common core State Standards.

The SBAC is a computer adaptive test. It means that there is a set of test questions in a variety of question types that adjust to each student based on the student's answers to previous questions. This section includes a range of items types, such as selecting several correct responses for one item, typing out a response, fill---in short answers/tables, graphing, drag and drop, etc.

On computer adaptive tests, if the correct answer is chosen, the next question will be harder. If the answer given is incorrect, the next question will be easier. This also means that once an answer is selected on the computer it cannot be changed.

In this section, there are 2 complete SBAC Math Tests that reflect the format and question types on SBAC. On a real SBAC Math test, the number of questions varies and there are about 30 questions.

Let your student take these tests to see what score he or she will be able to receive on a real SBAC test.

Time to Test

Time to refine your skill with a practice examination.

Take a REAL SBAC Mathematics test to simulate the test day experience. After you've finished, score your test using the answer key.

Before You Start

- You'll need a pencil and scratch papers to take the test.
- For this practice test, don't time yourself. Spend time as much as you need.
- It's okay to guess. You won't lose any points if you're wrong.
- After you've finished the test, review the answer key to see where you went wrong.

Calculators are not permitted for SBAC Tests.

Good Luck!

SBAC GRADE 6 MAHEMATICS REFRENCE MATERIALS

Conversions:

LENGTH	
Customary	**Metric**
1 mile (mi) = 1,760 yards (yd)	1 kilometer (km) = 1,000 meters (m)
1 yard (yd) = 3 feet (ft)	1 meter (m) = 100 centimeters (cm)
1 foot (ft) = 12 inches (in.)	1 centimeter (cm) = 10 millimeters (mm)

VOLUME AND CAPACITY	
Customary	**Metric**
1 gallon (gal) = 4 quarts (qt)	1 liter (L) = 1,000 milliliters (mL)
1 quart (qt) = 2 pints (pt.)	
1 pint (pt.) = 2 cups (c)	
1 cup (c) = 8 fluid ounces (Fl oz)	

WEIGHT AND MASS	
Customary	**Metric**
1 ton (T) = 2,000 pounds (lb.)	1 kilogram (kg) = 1,000 grams (g)
1 pound (lb.) = 16 ounces (oz)	1 gram (g) = 1,000 milligrams (mg)

Formulas:

Area	
Triangle	$A = \frac{1}{2}bh$
Rectangle or Parallelogram	$A = bh$
Trapezoid	$A = \frac{1}{2}h(b_1 + b_2)$
Volume	
Rectangular Prism	$V = Bh$

Smarter Balanced Assessment Consortium

SBAC Practice Test 1

Mathematics

GRADE 6

- 30 Questions
- There is no time limit for this practice test.
- Basic Calculators are permitted for this practice test.

Released Month Year

1) If $x = -7$, which of the following equations is true?

A. $x(2x + 3) = 75$

B. $x(35 - x^2) = 108$

C. $3(x^2 - 18) = 66$

D. $x(-4x - 3) = -175$

2) What is the perimeter of the following shape? (it's a right triangle)

A. 45 cm

B. 40 cm

C. 62 cm

D. 72 cm

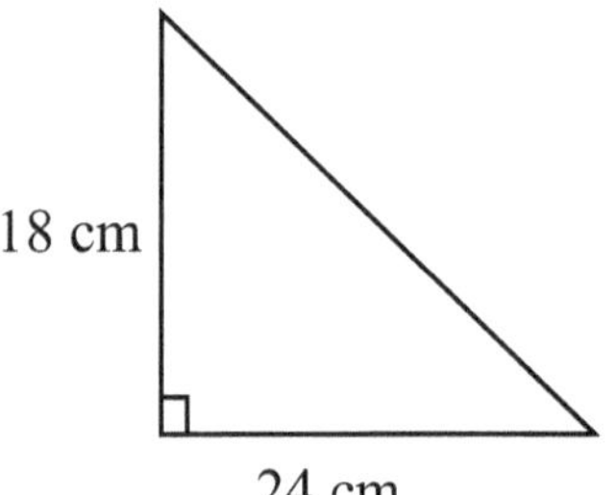

3) 72 is what percent of 15?

A. 120 %

B. 960 %

C. 240 %

D. 480 %

4) Which of the following expressions has a value of -24?

A. $-7 + (-45 \div 9) + \frac{-7}{8} \times 8$

B. $-8 \times (-11) + (-5) \times 5$

C. $(-6) + 12 \times 6 \div (-4)$

D. $(-7) \times (-3) + 5$

5) 564 inches equal to …?

A. 47 ft.

B. 210 ft.

C. 43 ft.

D. 95 ft.

6) Which of the following equations is true?

A. $0.06 = \frac{6}{10}$

B. $\frac{30}{100} = 0.03$

C. $5.8 = \frac{58}{10}$

D. $\frac{45}{9} = 0.5$

7) What is the greatest common factor of 18 and 90?

A. 18

B. 9

C. 90

D. 27

8) Which list shows the integer numbers listed in order from least to greatest?

A. $-37, -28, -16, 1, 11, 18$

B. $-16, -28, -37, 11, 18, 1$

C. $-37, -16, -28, 11, 18, 1$

D. $-28, -37, -16, 11, 1, 18$

9) Based on the table below, which of the following expressions represents any value of f in term of its corresponding value of x?

x	1.25	1.75	2.25
$f(x)$	2	2.5	3

A. $f(x) = 3x - \frac{3}{4}$

B. $f(x) = x + \frac{3}{4}$

C. $f(x) = 2x + \frac{1}{2}$

D. $f(x) = 2x - 1\frac{1}{2}$

10) A football team won exactly 70% of the games it played during last session. Which of the following could be the total number of games the team played last season?

A. 49

B. 40

C. 55

D. 77

11) The perimeter of the trapezoid below is 66. What is its area?

A. 420 cm^2

B. 210 cm^2

C. 164 cm^2

D. 105 cm^2

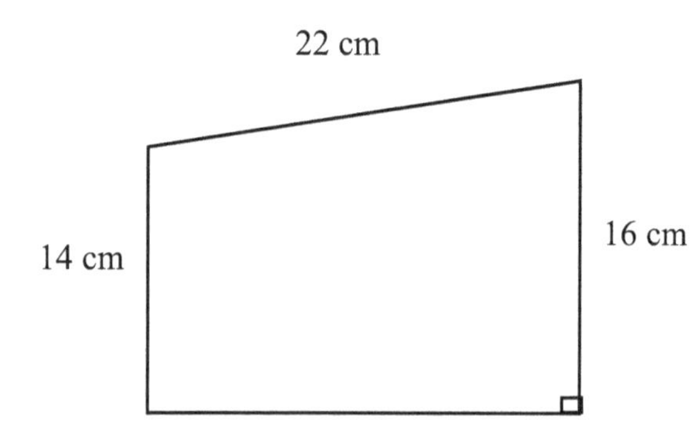

12) There are 77 blue marbles and 187 red marbles. We want to place these marbles in some boxes so that there is the same number of red marbles in each box and the same number of blue marbles in each of the boxes. How many boxes do we need?

A. 11

B. 7

C. 9

D. 19

13) Car A travels 198.6 km at a given time, while car B travels 1.9 times the distance car A travels at the same time. What is the distance car B travels during that time?

A. 307.74 km

B. 299.35 km

C. 298.54 km

D. 377.34 km

14) Which of the following expressions has the greatest value?

A. $12^2 - 3^4$

B. $4^4 - 6^3$

C. $7^3 - 16^2$

D. $5^3 - 10^2$

15) The diameter of a circle is 10π. What is the area of the circle?

A. $10\pi^2$

B. $\frac{100\pi^3}{2}$

C. $25\pi^3$

D. $\frac{10\pi^3}{4}$

16) Elise has x apples. Alvin has 47 apples, which is 25 apples less than number of apples Elise owns. If Baron has $\frac{1}{4}$ times as many apples as Elise has. How many apples does Baron have?

A. 16

B. 72

C. 36

D. 18

17) Find the perimeter of shape in the following figure? (all angles are right angles)

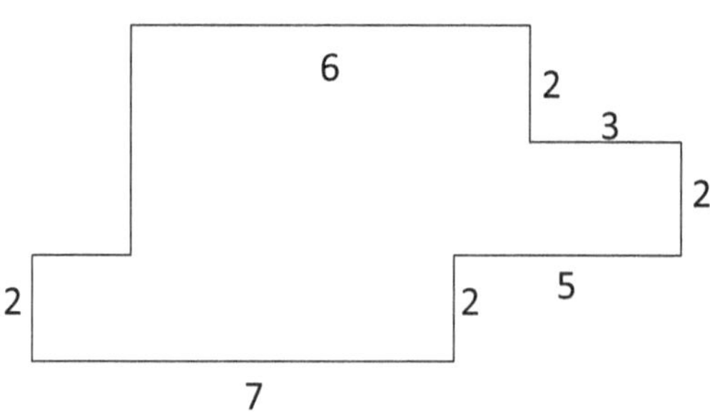

A. 42

B. 36

C. 48

D. 24

18) In the following triangle find α.

A. 96°

B. 63°

C. 64°

D. 47°

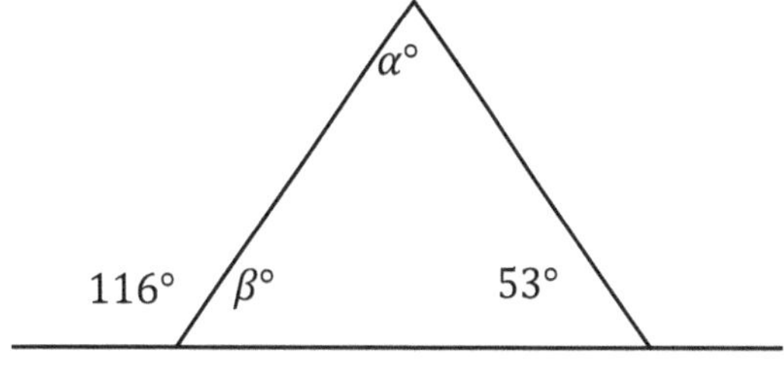

19) What is the probability of choosing a month starts with M in a year?

A. $\frac{1}{2}$

B. 1

C. $\frac{1}{3}$

D. $\frac{1}{6}$

20) $5(1.3) - 3.95 = \cdots ?$

A. 2.65

B. 2.55

C. 3.65

D. 3.55

21) If point A placed at $-\frac{45}{9}$ on a number line, which of the following points has a distance equal to 4 from point A?

A. -9

B. -1

C. 4

D. A and B

22) What are the values of mode and median in the following set of numbers?

$$7, 1, 7, 4, 4, 2, 2, 3, 7, 7, 3$$

A. Mode: 4, Median: 3

B. Mode: 7, Median: 7

C. Mode: 3, Median: 4

D. Mode: 7, Median: 4

23) The ratio of pens to pencils in a box is 5 to 8. If there are 208 pens and pencils in the box altogether, how many more pens should be put in the box to make the ratio of pens to pencils 1: 1?

A. 80

B. 48

C. 128

D. 64

24) Which of the following shows the numbers in increasing order?

A. $\frac{6}{5}, \frac{8}{3}, \frac{7}{12}, \frac{26}{4}$

B. $\frac{7}{12}, \frac{8}{3}, \frac{6}{5}, \frac{26}{4}$

C. $\frac{7}{12}, \frac{6}{5}, \frac{8}{3}, \frac{26}{4}$

D. $\frac{26}{4}, \frac{8}{3}, \frac{7}{12}, \frac{6}{5}$

25) If $8x - 3 = 21$, what is the value of $4x + 7$?

A. 23

B. 19

C. 27

D. 15

Types of air pollutions in 10 cities of a country

Type of Pollution	Number of Cities									
A	■	■	■	■						
B	■	■	■	■	■	■	■	■		
C	■	■	■							
D	■	■	■	■	■	■	■			
E	■	■	■	■	■					
	1	**2**	**3**	**4**	**5**	**6**	**7**	**8**	**9**	**10**

26) Based on the above data, what percent of cities are in the type of pollution A, B, and D respectively?

A. 40%, 80%, 70%

B. 40%, 70%, 80%

C. 70%, 40%, 80%

D. 80%, 40%, 70%

27) How many tiles of 7 cm^2 is needed to cover a floor of dimension 3 cm by 49 cm?

A. 56

B. 28

C. 14

D. 21

28) A shaft rotates 250 times in 6 seconds. How many times does it rotate in 15 seconds?

A. 555

B. 625

C. 225

D. 565

29) Which of the following statement can describe the following inequality correctly?

$$\frac{x}{7} \geq 15$$

A. David put x books in 7 shelves, and each shelf had at least 15 books.

B. David placed 7 books in x shelves so that each shelf had less than 15 books.

C. David put 15 books in x shelves and each shelf had exactly 7 books.

D. David put x books in 7 shelves, and each shelf had more than 15 books.

30) Removing which of the following numbers will change the average of the numbers to 8.8?

3, 7, 14, 16, 11, 9

A. 16

B. 11

C. 9

D. 7

Smarter Balanced Assessment Consortium

SBAC Practice Test 2

Mathematics

GRADE 6

- **30 Questions**
- **There is no time limit for this practice test.**
- **Basic Calculators are permitted for this practice test.**

Released Month Year

1) Martin earns $19 an hour. Which of the following inequalities represents the amount of time Martin needs to work per day to earn at least $195 per day?

A. $19t \geq 195$

B. $19t \leq 195$

C. $19 + t \geq 195$

D. $19 + t \leq 195$

2) What is the value of the expression $3(4x - y) + (8 - 2x)^2$, when $x = 3$ and $y = -4$?

A. 48

B. -52

C. 52

D. -48

3) Round $\frac{216}{7}$ to the nearest tenth.

A. 31.9

B. 31.6

C. 30.9

D. 30.6

4) Find the opposite of the numbers 18, 7.

A. $\frac{1}{18}, 7$

B. $-18, 7$

C. $-18, -7$

D. $-\frac{1}{18}, 7$

5) Which ordered pair describes point A that is shown below?

A. $(1, -3)$

B. $(-1, 3)$

C. $(-1, 3)$

D. $(-1, -3)$

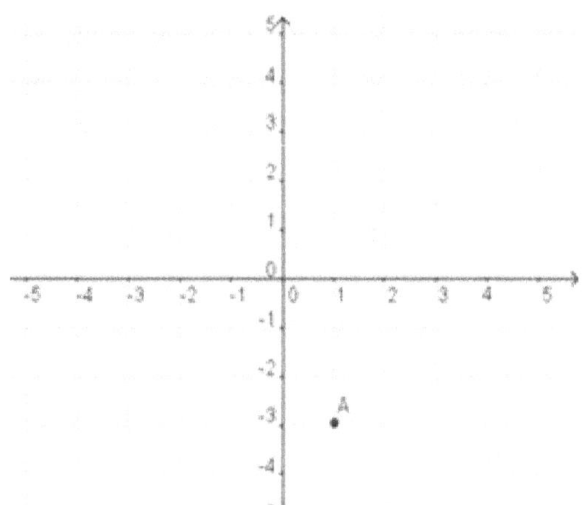

6) To produce a special concrete, for every 12 kg of cement, 4 liters of water is required. Which of the following ratios is the same as the ratio of cement to liters of water?

A. 48: 16

B. 48: 20

C. 36: 16

D. 60: 16

7) $(74 + 26) \div 27$ is equivalent to …

A. $100 \div 2.7$

B. $\frac{74}{27} + 6$

C. $(2 \times 2 \times 5 \times 5) \div (3 \times 3 \times 3)$

D. $(2 \times 2 \times 5 \times 5) \div 3 + 3 + 3$

8) Which of the following graphs represents the following inequality?

$$-5 \leq 3x - 8 < 7$$

A.

−8 −6 −4 −2 0 2 4 6 8

B.

−8 −6 −4 −2 0 2 4 6 8

C.

−8 −6 −4 −2 0 2 4 6 8

D.

−8 −6 −4 −2 0 2 4 6 8

9) The ratio of boys to girls in a school is 3:5. If there are 1,600 students in the school, how many boys are in the school?

A. 280

B. 820

C. 560

D. 600

10) What is the equation of a line that passes through points (1, 3) and (2, 8)?

A. $y = 5x$

B. $y = 5x - 2$

C. $y = 2x + 5$

D. $y = 7x - 5$

11) What is the volume of a box with the following dimensions? Height = 7cm

Width = 6 cm Length = 3 cm

A. 62 cm^3

B. 265 cm^3

C. 116 cm^3

D. 126 cm^3

12) Anita's trick–or–treat bag contains 11 pieces of chocolate, 14 suckers, 8 pieces of gum, 9 pieces of licorice. If she randomly pulls a piece of candy from her bag, what is the probability of her pulling out a piece of sucker?

A. $\frac{1}{5}$

B. $\frac{1}{3}$

C. $\frac{2}{5}$

D. $1 = \frac{14}{14}$

13) Which statement is true about all square?

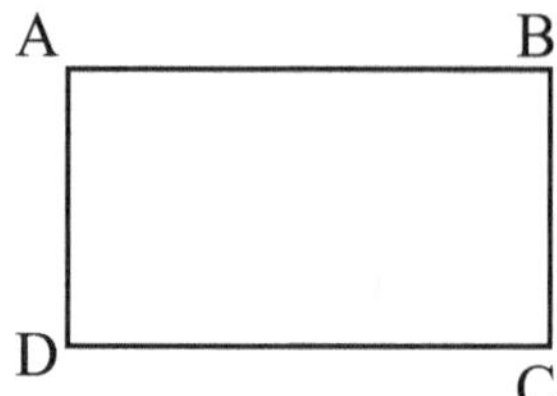

A. Both diagonals have equal measure.

B. All sides are congruent.

C. Both diagonals are perpendicular.

D. All the statements are true.

14) Which of the following lists shows the fractions in order from least to greatest?

$$\frac{16}{15}, \frac{4}{5}, \frac{3}{7}, \frac{19}{6}$$

A. $\frac{19}{6}, \frac{4}{5}, \frac{3}{7}, \frac{16}{15}$

B. $\frac{4}{5}, \frac{16}{15}, \frac{19}{6}, \frac{3}{7}$

C. $\frac{3}{7}, \frac{4}{5}, \frac{16}{15}, \frac{19}{6}$

D. $\frac{16}{15}, \frac{4}{5}, \frac{19}{6}, \frac{3}{7}$

15) Which statement about 7 multiplied by $\frac{7}{9}$ must be true?

A. The product is between 4.2 and 5.

B. The product is greater than 5.

C. The product is equal to $\frac{87}{32}$.

D. The product is between 2 and 3.

16) A car costing $500 is discounted 15%. Which of the following expressions can be used to find the selling price of the car?

A. $(500)(0.05)$

B. $500 - (500 \times 0.15)$

C. $(500)(0.15)$

D. $500 - (500 \times 0.85)$

17) The distance between two cities is 5,271 feet. What is the distance of the two cities in yards?

A. 1,757 yd.

B. 2,500 yd.

C. 3,775 yd.

D. 2,557 yd.

18) 78 is equal to …

A. $-15 - (5 \times 12) + (6 \times 21)$

B. $\left(\frac{21}{7} \times 56\right) - \left(\frac{222}{3}\right)$

C. $\left(\left(\frac{22}{7} + \frac{16}{7}\right) \times 14\right) - \frac{54}{6} + \frac{132}{12}$

D. $\frac{460}{14} + \frac{232}{7} - 19$

19) Mr. Jones saves \$2,500 out of his monthly family income of \$60,000. What fractional part of his income does Mr. Jones save?

A. $\frac{1}{24}$

B. $\frac{1}{6}$

C. $\frac{5}{6}$

D. $\frac{1}{48}$

20) If the area of the following trapezoid is equal to A, which equation represent x?

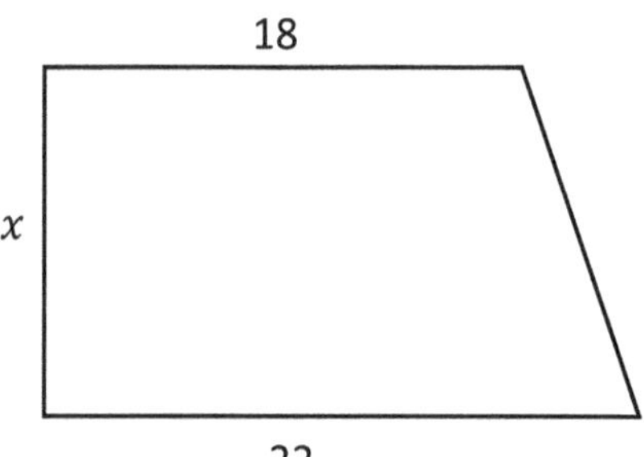

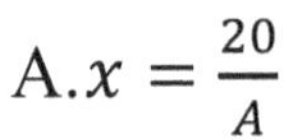
A. $x = \frac{20}{A}$

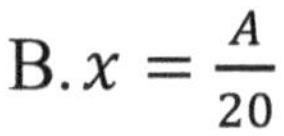
B. $x = \frac{A}{20}$

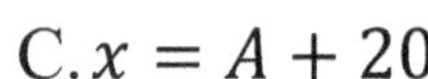
C. $x = A + 20$

D. $x = A - 20$

21) By what factor did the number below change from first to fourth number?

13, 78, 468, 2808

A. 6

B. 12

C. 24

D. 62

22) Based on the table below, which expression represents any value of f in term of its corresponding value of x?

x	1.3	2.2	3.5
$f(x)$	6.3	9.9	15.1

A. $f(x) = 5x - 2\frac{1}{5}$

B. $f(x) = 5x + 2\frac{1}{5}$

C. $f(x) = 4x + 1\frac{1}{10}$

D. $f(x) = 3x + \frac{3}{5}$

23) Calculate the approximate area of the following circle? (the diameter is 11)

A. 56

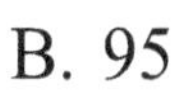
B. 95

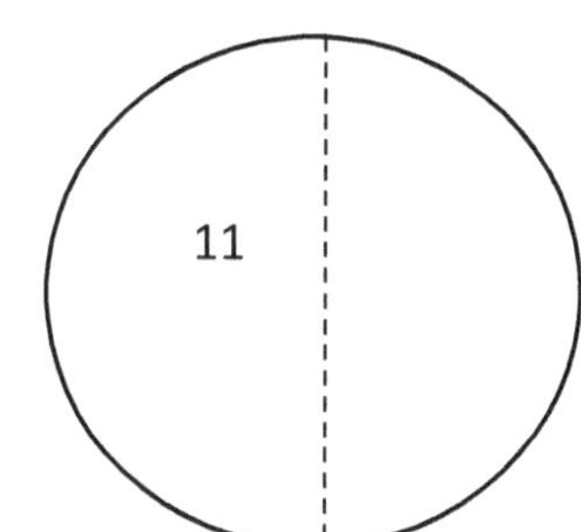

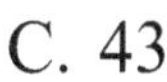
C. 43

D. 46.6

24) The following graph shows the mark of seven students in mathematics. What is the mean (average) of the marks?

A. 14.45

B. 13.21

C. 12.54

D. 14.64

25) Which of the following statements is correct, according to the graph below?

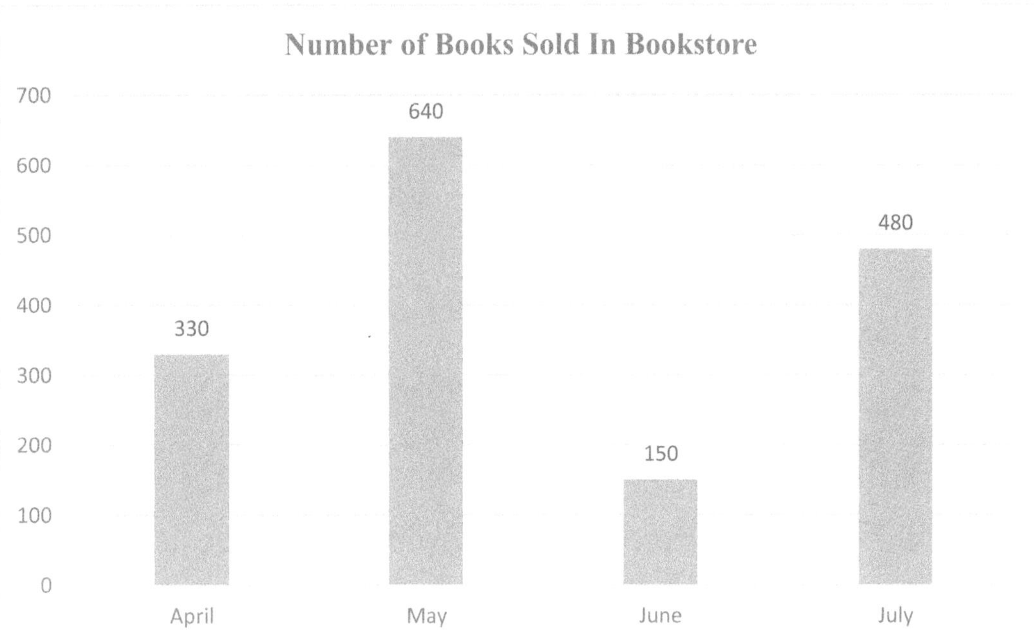

A. The number of books sold in the April was twice the number of books sold in the July.

B. The number of books sold in the July was less than one fourth the number of books sold in the May.

C. The number of books sold in the June was more than one-third the number of books sold in the April.

D. The number of books sold in the May was equal to the number of books sold in April plus the number of books sold in the June.

26) If the area of the following rectangular ABCD is 60, and E is the midpoint of AB, what is the area of the shaded part?

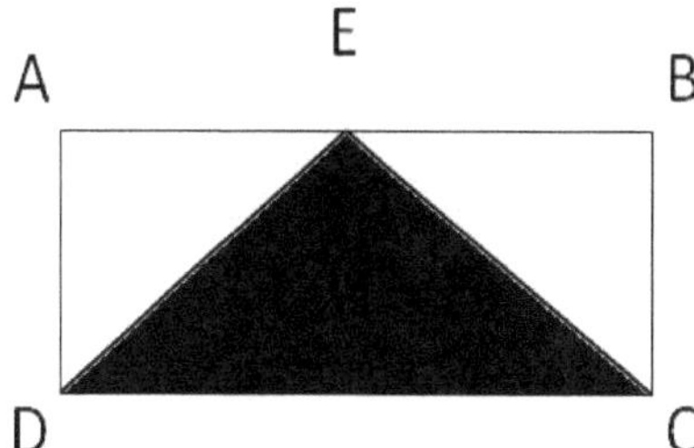

A. 40

B. 30

C. 120

D. 75

27) What is the ratio between α and β $\left(\frac{\alpha}{\beta}\right)$ in the following shape?

A. $\frac{5}{6}$

B. $\frac{1}{3}$

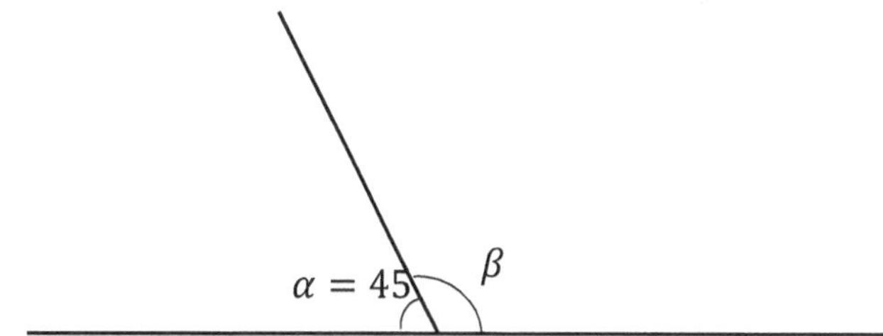

C. $\frac{3}{4}$

D. $\frac{1}{6}$

28) When point A $(-9, 4)$ is reflected over the y-axis to get the point B, what are the coordinates of point B?

A. $(-9, -4)$

B. $(9, -4)$

C. $(9, 4)$

D. $(-9, 4)$

29) An integer is chosen at random from 1 to 20. Find the probability of not selecting a composite number?

A. $\frac{11}{20}$

B. $\frac{7}{20}$

C. $\frac{9}{20}$

D. $\frac{3}{5}$

30) What is the lowest common multiple of 15 and 45?

A. 45

B. 15

C. 30

D. 5

Chapter 13 : Answers and Explanations

SBAC Practice Tests

Answer Key

❋ Now, it's time to review your results to see where you went wrong and what areas you need to improve!

Practice Test - 1					
1	D	11	B	21	D
2	D	12	A	22	D
3	D	13	D	23	B
4	C	14	C	24	C
5	A	15	C	25	B
6	C	16	D	26	A
7	A	17	B	27	D
8	A	18	B	28	B
9	B	19	D	29	A
10	B	20	B	30	A

Practice Test - 2					
1	A	11	D	21	A
2	C	12	B	22	C
3	C	13	D	23	B
4	C	14	C	24	B
5	A	15	B	25	C
6	A	16	B	26	B
7	C	17	A	27	B
8	C	18	C	28	C
9	D	19	A	29	C
10	B	20	B	30	A

Practice Test 1

SBAC - Mathematics

Answers and Explanations

1) Answer: D

Plugin the value of x in the equations. $x = -7$, then:

A. $x(2x+3) = 75 \rightarrow (-7)(2(-7)+3) = (-7)(-11) = 77 \neq 75$

B. $x(35 - x^2) = 108 \rightarrow (-7)(35 - (-7)^2) = (-7)(35 - 49) = (-7)(-14) = 98 \neq 108$

C. $3(x^2 - 18) = 66 \rightarrow 3\left((-7)^2 - 18\right) = 3(49 - 18) = 3(31) = 93 \neq 66$

D. $x(-4x - 3) = -175 \rightarrow -7(-4(-7) - 3) = -7(28 - 3) = -175 = -175$

2) Answer: D

Use Pythagorean theorem to find the hypotenuse of the triangle.

$$a^2 + b^2 = c^2 \rightarrow 18^2 + 24^2 = c^2 \rightarrow 324 + 576 = c^2 \rightarrow 900 = c^2 \rightarrow c = 30$$

The perimeter of the triangle is: $18 + 24 + 30 = 72$

3) Answer: D

Use percent formula: Part $= \frac{\text{percent}}{100} \times$ whole.

$72 = \frac{\text{percent}}{100} \times 15 \Rightarrow 72 = \frac{\text{percent} \times 15}{100}$

$\Rightarrow 72 = \frac{\text{percent} \times 3}{20}$, multiply both sides by 20.

$1{,}440 = \text{percent} \times 3$, divide both sides by 3.

$480 = \text{percent}$; The answer is 480%

4) Answer: C

Let's check the options provided.

A. $-7 + (-45 \div 9) + \frac{-7}{8} \times 8 \rightarrow -7 + (-5) + (-7) = -19$

B. $-8 \times (-11) + (-5) \times 5 = (88) + (-25) = 63$

C. $(-6) + 12 \times 6 \div (-4) = -6 + 72 \div (-4) = -6 - 18 = -24$

D. $(-7) \times (-3) + 5 = 21 + 5 = 26$

5) Answer: A

1 feet = 12 inches. Then: $564 \text{ in } \times \frac{1 \text{ ft}}{12 \text{ in}} = \frac{564}{12} \text{ ft} = 47 \text{ ft}$

6) Answer: C

A. $0.06 = \frac{6}{100}$

B. $\frac{30}{100} = \frac{3}{10} = 0.3$

C. $5.8 = 5\frac{8}{10} = \frac{58}{10}$

D. $\frac{45}{9} = 5$

7) Answer: A

Prime factorizing of $18 = 2 \times 3 \times 3$

Prime factorizing of $90 = 2 \times 3 \times 3 \times 5$

To find Greatest Common Factor, multiply the common factors of both numbers.

GCF= $2 \times 3 \times 3 = 18$

8) Answer: A

$-37 < -28 < -16 < 1 < 11 < 18$

Then choice A

9) Answer: B

Plug in the values of x in the equations provided.

A. $f(x) = 3x - \frac{3}{4} = 3(1.25) - \frac{3}{4} = 3.75 - 0.75 = 3 \neq 2$

B. $f(x) = x + \frac{3}{4} = 1.25 + \frac{3}{4} = 1.25 + 0.75 = 2$

C. $f(x) = 2x + \frac{1}{2} = 2(1.25) + \frac{1}{2} = 2.5 + 0.5 = 3 \neq 2$

D. $f(x) = 2x - 1\frac{1}{2} = 2(1.25) - \frac{3}{2} = 2.5 - 1.5 = 1 \neq 2$

10) Answer: B

Choices A, C and D are incorrect because 70% of each of the numbers is a non-whole number.

A. 49, $70\% \; of \; 49 = 0.70 \times 49 = 34.3$

B. 40, $70\% \; of \; 40 = 0.70 \times 40 = 28$

C. 55, $70\%\ of\ 55 = 0.70 \times 55 = 38.5$

D. 77, $70\%\ of\ 77 = 0.70 \times 77 = 53.9$

11) Answer: B

The perimeter of the trapezoid is 66.

Therefore, the missing side (height) is = 66 – 14 – 16 – 22 = 14

Area of the trapezoid: $A = \frac{1}{2}h(b1 + b2) = \frac{1}{2}(14)\ (14 + 16)\ =\ 210$

12) Answer: A

First, we need to find the GCF (Greatest Common Factor) of 187 and 77.

$187 = 17 \times 11$

$77 = 7 \times 11 \rightarrow$ GCF = 11; Therefore, we need 11 boxes.

13) Answer: D

Distance that car B travels = $1.9 \times$ distance that car A travels

$= 1.9 \times 198.6 = 377.34$ Km

14) Answer: C

A. $12^2 - 3^4 = 144 - 81 = 63$

B. $4^4 - 6^3 = 256 - 216 = 40$

C. $7^3 - 16^2 = 343 - 256 = 87$

D. $5^3 - 10^2 = 125 - 100 = 25$

15) Answer: C

The radius of the circle is: $\frac{10\pi}{2} = 5\pi$

The area of circle: $\pi r^2 = \pi(5\pi)^2 = \pi \times 25\pi^2 = 25\pi^3$

16) Answer: D

Elise has x apple which is 25 apples more than number of apples Alvin owns.

Therefore:

$x - 25 = 47 \rightarrow x = 47 + 25 = 72$

Elise has 72 apples.

Let y be the number of apples that Baron has. Then: $y = \frac{1}{4} \times 72 = 18$

17) Answer: B

Let x and y be two sides of the shape. Then:

$x + 2 = 2 + 2 + 2 \rightarrow x = 4$

$y + 6 + 3 = 7 + 5 \rightarrow y + 9 = 12 \rightarrow y = 3$

Then, the perimeter is:

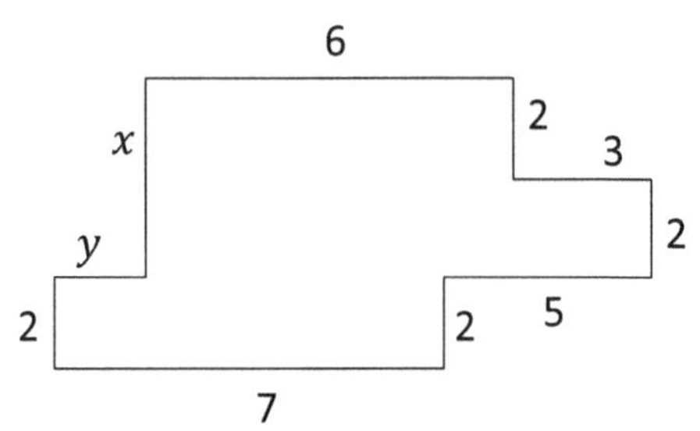

$2 + 7 + 2 + 5 + 2 + 3 + 2 + 6 + 4 + 3 = 36$

18) Answer: B

Supplementary angles add up to 180 degrees.

$\beta + 116° = 180° \rightarrow \beta = 180° - 116° = 64°$

The sum of all angles in a triangle is 180 degrees. Then:

$\alpha + \beta + 53° = 180° \rightarrow \alpha + 64° + 53° = 180°$

$\rightarrow \alpha + 117° = 180° \rightarrow \alpha = 180° - 117° = 63°$

19) Answer: D

Two months, March and May, in 12 months start with M, then:

Probability $= \frac{\textit{number of desired outcomes}}{\text{number of total outcomes}} = \frac{2}{12} = \frac{1}{6}$

20) Answer: B

$5(1.3) - 3.95 = 6.50 - 3.95 = 2.55$

21) Answer: D

If the value of point A is greater than the value of point B, then the distance of two points on the number line is: value of A− value of B.

A. $-\frac{45}{9} - (-9) = -5 + 9 = 4 = 4$

B. $-1 - \left(-\frac{45}{9}\right) = -1 + 5 = 4 = 4$

C. $4 - \left(-\frac{45}{9}\right) = 4 + 5 = 9 \neq 4$

22) Answer: D

First, put the numbers in order from least to greatest: 1, 2, 2, 3, 3, 4, 4, 7, 7, 7, 7

The Mode of the set of numbers is: 7 (the most frequent numbers)

Median is: 4 (the number in the middle)

23) Answer: B

The ratio of pens to pencils is 5: 8. Therefore there are 5 pens out of all 13 pens and pencils. To find the answer, first dived 208 by 13 then multiply the result by 5: $208 \div 13 = 16 \rightarrow 16 \times 5 = 80$

There are 80 pens and 128 pencils (208− 80). Therefore, 48 more pens should be put in the box to make the ratio 1: 1.

24) Answer: C

$\frac{26}{4} = 6.5$ $\frac{6}{5} = 1.2$ $\frac{7}{12} \cong 0.58$ $\frac{8}{3} \cong 2.67$

Then: $\frac{7}{12} < \frac{6}{5} < \frac{8}{3} < \frac{26}{4}$

25) Answer: B

$8x - 3 = 21 \rightarrow 8x = 21 + 3 = 24 \rightarrow x = \frac{24}{8} = 3$

Then, $4x + 7 = 4(3) + 7 = 12 + 7 = 19$

26) Answer: A

Percent of cities in the type of pollution A: $\frac{4}{10} \times 100 = 40\%$

Percent of cities in the type of pollution B: $\frac{8}{10} \times 100 = 80\%$

Percent of cities in the type of pollution D: $\frac{7}{10} \times 100 = 70\%$

27) Answer: D

The area of the floor is: 3 cm × 49 cm = 147 cm

The number of tiles needed = 147 ÷7 = 21

28) Answer: B

The shaft rotates 250 times in 6 seconds. Then, the number of rotates in 15 second equals to: $\frac{250 \times 15}{6} = 625$

29) Answer: A

Let's write an inequality for each statement.

A. $\frac{x}{7} \geq 15$ (this is the same as the inequality provided)

B. $\frac{7}{x} < 15$

C. $\frac{15}{x} = 7$

D. $\frac{x}{7} > 15$

30)Answer: A

Check each option provided:

A. 16 $\frac{3+7+14+11+9}{5} = \frac{44}{5} = 8.8$

B. 11 $\frac{3+7+14+16+9}{5} = \frac{49}{5} = 9.8$

C. 9 $\frac{3+7+14+16+11}{5} = \frac{51}{5} = 10.2$

D. 7 $\frac{3+14+11+16+9}{5} = \frac{53}{5} = 10.6$

Practice Test 2

SBAC - Mathematics

Answers and Explanations

1) Answer: A

For one hour he earns \$19, then for t hours he earns \$19t. If he wants to earn at least \$195, therefor, the number of working hours multiplied by 19 must be equal to 195 or more than 195. $19t \geq 195$

2) Answer: C

Plug in the value of x and y and use order of operations rule. $x = 3$ and $y = -4$

$3(4x - y) + (8 - 2x)^2 = 3(4(3) - (-4)) + (8 - 2(3))^2 = 3(12 + 4) + (2)^2 =$

$48 + 4 = 52$

3) Answer: C

$\frac{216}{7} \cong 30.85 \cong 30.9$

4) Answer: C

Opposite number of any number x is a number that if added to x, the result is 0. Then:

$18 + (-18) = 0$ and $7 + (-7) = 0$

5) Answer: A

The coordinate plane has two axes. The vertical line is called the y-axis and the horizontal is called the x-axis. The points on the coordinate plane are address using the form (x, y). The point A is one unit on the right side of x-axis; therefore, its x value is 1 and it is three unit down, therefore its y axis is -3. The coordinate of the point is: $(1, -3)$

6) Answer: A

48: 16 = 12: 4

$12 \times 4 = 48$ And $4 \times 4 = 16$

7) Answer: C

$(74 + 26) \div (27) = (100) \div (27)$

The prime factorization of 100 is: $2 \times 2 \times 5 \times 5$

The prime factorization of 27 is: $3 \times 3 \times 3$

Therefore: $(100) \div (27) = (2 \times 2 \times 5 \times 5) \div (3 \times 3 \times 3)$

8) Answer: C

Solve for x.

$-5 \leq 3x - 8 < 7 \Rightarrow$ (add 8 all sides) $-5 + 8 \leq 3x - 8 + 8 < 7 + 8 \Rightarrow$

$3 \leq 3x < 15 \Rightarrow$ (divide all sides by 3)$1 \leq x < 5$

x is between 1 and 5. Choice C represent this inequality.

9) Answer: D

The ratio of boy to girls is 3:5. Therefore, there are 3 boys out of 8 students. To find the answer, first divide the total number of students by 8, then multiply the result by 3.

$1{,}600 \div 8 = 200 \Rightarrow 200 \times 3 = 600$

10) Answer: B

The slope of the line is: $\frac{y_2 - y_1}{x_2 - x_1} = \frac{8-3}{2-1} = \frac{5}{1} = 5$

The equation of a line can be written as:

$y - y_0 = m(x - x_0) \rightarrow y - 3 = 5(x - 1) \rightarrow y - 3 = 5x - 5 \rightarrow y = 5x - 2$

11) Answer: D

Volume of a box = length × width × height = $3 \times 6 \times 7 = 126$

12) Answer: B

Probability = $\frac{\textit{number of desired outcomes}}{\text{number of total outcomes}} = \frac{14}{11+14+8+9} = \frac{14}{42} = \frac{1}{3}$

13) Answer: D

In any square, all the statements are true.

14) Answer: C

Let's compare each fraction: $\frac{3}{7} < \frac{4}{5} < \frac{16}{15} < \frac{19}{6}$

Only choice C provides the right order.

15) Answer: B

$7 \times \frac{7}{9} = \frac{49}{9} = 5.44$

A. $5.44 > 5$

B. $5.44 > 5$ This is the answer!

C. $\frac{87}{32} = 2.72 \neq 5.44$

D. $3 < 5.44$

16) Answer: B

To find the discount, multiply the number $(100\% - \text{rate of discount})$

Therefore; $500(100\% - 15\%) = 500(1 - 0.15) = 500 - (500 \times 0.15)$

17) Answer: A

1 yard = 3 feet

Therefore, 5,271 ft.$\times \frac{1\text{ yd}}{3\text{ ft}} = 1{,}757$ yd

18) Answer: C

Simplify each option provided.

A. $-15 - (5 \times 12) + (6 \times 21) = -15 - 60 + 126 = 51$

B. $\left(\frac{21}{7} \times 56\right) - \left(\frac{222}{3}\right) = 168 - 74 = 94$

C. $\left(\left(\frac{22}{7} + \frac{16}{7}\right) \times 14\right) - \frac{54}{6} + \frac{132}{12} = \left(\left(\frac{22+16}{7}\right) \times 14\right) - \frac{54}{6} + \frac{132}{12} = \left(\left(\frac{38}{7}\right) \times 14\right) +$

$\frac{132-108}{12} = (38 \times 2) + \frac{24}{12} = 76 + 2 = 78$ (this is the answer)

D. $\frac{460}{14} + \frac{232}{7} - 19 = \frac{460+464}{14} - 19 = 66 - 19 = 47$

19) Answer: A

2,500 out of 60,000 equals to $\frac{2{,}500}{60{,}000} = \frac{1}{24}$

20) Answer: B

The area of the trapezoid is: $area = \frac{(base\ 1 + base\ 2)}{2} \times height = \left(\frac{18+22}{2}\right)x = A \rightarrow$

$20x = A \rightarrow x = \frac{A}{20}$

21) Answer: A

$\frac{78}{13} = 6, \frac{468}{78} = 6, \frac{2{,}808}{468} = 6$

Therefore, the factor is 6

22) Answer: C

Plug in the value of x into the function $f(x)$. First, plug in 1.3 for x.

A. $f(x) = 5x - 2\frac{1}{5} = 5(1.3) - 2\frac{1}{5} = 4.3 \neq 6.30$

B. $f(x) = 5x + 2\frac{1}{5} = 5(1.3) + 2\frac{1}{5} = 8.7 \neq 6.30$

C. $f(x) = 4x + 1\frac{1}{10} = 4(1.3) + 1\frac{1}{10} = 5.2 + 1.1 = 6.3$ This is correct!

Plug in other values of x. $x = 2.2$

$f(x) = 4x + 1\frac{1}{10} = 4(2.2) + 1.1 = 9.9$ This one is also correct. $x = 3.5$

$f(x) = 4x + 1\frac{1}{10} = 4(3.5) + 1.1 = 15.1$ This one works too!

D. $3x + \frac{3}{5} = 3(1.30) + \frac{3}{5} = 4.5 \neq 6.30$

23) Answer: B

The diameter of a circle is twice the radius. Radius of the circle is $\frac{11}{2}$.

Area of a circle $= \pi r^2 = \pi(\frac{11}{2})^2 = 30.25\pi = 30.25 \times 3.14 = 94.98 \cong 95$

24) Answer: B

Average (mean) $= \frac{\text{sum of terms}}{\text{number of terms}} = \frac{12+13+15+11+13+17+11.5}{7} = 13.21$

25) Answer: C

A. Number of books sold in April is: 330

Number of books sold in July is:480 $\rightarrow \frac{330}{480} = \frac{11}{16} \neq 2$

B. Number of books sold in July is: 480

One fourth the number of books sold in May is: $\frac{640}{4} = 160 \rightarrow 480 > 160$

C. Number of books sold in June is: 150

One-third the number of books sold in April is: $\frac{330}{3} = 110 \rightarrow 150 > 110$ (it's correct)

D. $330 + 150 = 480 > 640$

26) Answer: B

Since, E is the midpoint of AB, then the area of all triangles DAE, DEF, CFE and CBE are equal.

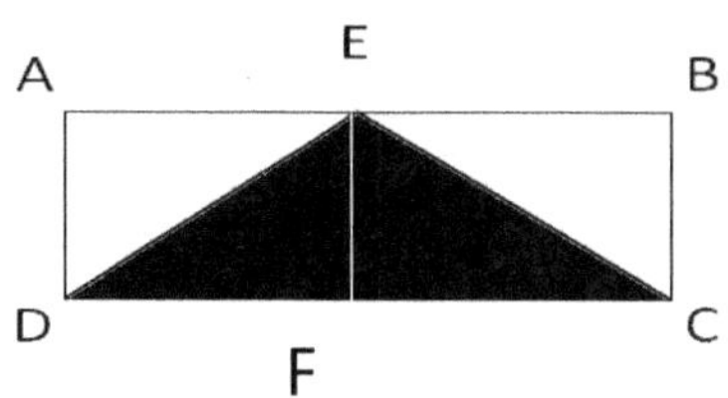

Let x be the area of one of the triangles, then:

$4x = 60 \rightarrow x = 15$

The area of DEC $= 2x = 2(15) = 30$

27) Answer: B

α and β are supplementary angles. The sum of supplementary angles is 180 degrees.

$\alpha + \beta = 180° \rightarrow \beta = 180° - \alpha = 180° - 45° = 135°$

Then, $\frac{\alpha}{\beta} = \frac{45}{135} = \frac{1}{3}$

28) Answer: C

When points are reflected over y-axis, the value of y in the coordinates doesn't change and the sign of x changes. Therefore, the coordinates of point B are $(9, 4)$.

29) Answer: C

There are 20 integers from 1 to 20. Set of numbers that are not composite between 1 and 20 is: A= {1, 2, 3, 5, 7, 11, 13, 17, 19}

9 integers are not composite. Probability of not selecting a composite number is:

Probability $= \frac{number\ of\ desired\ outcomes}{\text{number of total outcomes}} = \frac{9}{20}$

30) Answer: A

Prime factorizing of $45 = 3 \times 3 \times 5$

Prime factorizing of $15 = 3 \times 5$

LCM$=3 \times 3 \times 5 = 45$

"End"

www.ingramcontent.com/pod-product-compliance
Lightning Source LLC
LaVergne TN
LVHW061246100826
845148LV00008B/1041
* 9 7 8 1 6 3 6 2 0 0 7 3 6 *